Lecture Notes in Mathematics

Edited by A. Dold and B. Eckmann

Subseries: Fondazione C.I.M.E., Firenze

Adviser: Roberto Conti

1047

Fluid Dynamics

Lectures given at the 3rd 1982 Session of the Centro Internazionale Matematico Estivo (C.I.M.E.) held at Varenna, Italy, August 22 – September 1, 1982

Edited by H. Beirão da Veiga

Springer-Verlag
Berlin Heidelberg New York Tokyo 1984

Editor

H. Beirão da Veiga
Dipartimento di Matematica, Università
38050 Povo (Trento), Italy

AMS Subject Classifications (1980): 35D05, 35F25, 35F30, 35L60, 76N10

ISBN 3-540-12893-X Springer-Verlag Berlin Heidelberg New York Tokyo
ISBN 0-387-12893-X Springer-Verlag New York Heidelberg Berlin Tokyo

Printing and binding: Beltz Offsetdruck, Hemsbach/Bergstr.
2146/3140-543210

INTRODUCTION

This volume contains lectures and seminars presented at the "International Mathematical Summer Center" on "Fluid Dynamics" held at the "Villa Monastero", Varenna, Italy from August 22 to September 1, 1982.

The Session was organized by "Fondazione CIME" and sponsored by the Consiglio Nazionale delle Ricerche.

In organizing the meeting we have attempted to bring together different aspects in the field of fluid dynamics. Morning sessions were occupied by survey talks given by Prof. C. Bardos (Université Paris-Nord, France), Prof. A. Majda (University of California, Berkeley, U.S.A.) and Prof. J. Serrin (University of Minnesota, Minneapolis, U.S.A.).

The first part of this proceedings consists of C. Bardos and A. Majda lectures. Prof. Serrin's lectures on "Concepts of Continuum Thermo-mechanics" will appear later as part of a book. The second part consists of the afternoon seminar meetings about recent progress on the field.

I wish to express my thanks to the main lecturers, to all active participants who supported the meeting by their stimulating discussions and to the CIME scientific committee for the invitation to organize the conference.

Trento, October 1983.

Hugo BEIRÃO DA VEIGA

C.I.M.E. Session on Fluid Dynamics

List of participants

M. Arai, Emmyoji-Kitaura 2-13 (11-404), Ooyamazaki, Kyoto 618

M. Asadzadeh, Chalmers University of Technology, Department of Mathematics, S-412 96 Goteborg

J. Audounet, Université P. Sabatier, 118 Route de Narbonne, 31077 Toulouse

F.Bampi, Istituto Matematico Università, Via L.B. Alberti 4, 16132 Genova

I. Barbieri, Istituto Matematico Università, Piazza di Porta S. Donato 5, 40127 Bologna

C. Bardos, Université Paris-Nord, Centre Scientifique et Polytechnique, Av. J. B. Clement, 93430 Villetaneuse

E. Batta, Via Del Cantone 6, 06100 Perugia

H. Beirão da Veiga, Dipartimento di Matematica, Università, 38050 Povo (Trento)

M. Benati, Istituto Matematico Università, Via L.B. Alberti 4, 16132 Genova

P. Cannarsa, Via O. Tommasini 34, 00162 Roma

A. Ceré, Via G. B. Cortesi 6, 40141 Bologna

F. Conti, Scuola Normale Superiore, 56100 Pisa

T. Elmroth, Matematiska Institutionen, Chalmers Tekniska Hogskola, S-412 96 Goteborg

M. Fabrizio, Via S. Frediano 10, 40136 Bologna

A. Fernandez, Catedra de Mecanica de Fluidos, E.T.S. Ingenieros Industriales, Av. Reina Mercedes s/n, Sevilla 12

I. Ferrari, Via Ortensie 11, 41100 Modena

J. Fleckinger, 41 rue Boyssonne, 31400 Toulouse

F. Franchi, Via E. Toti 36, 40051 Altedo (Bologna)

L. Gardini, Viale della Repubblica 8, 40100 Bologna

G. Geymonat, Istituto Matematico del Politecnico, Corso Duca degli Abruzzi 24, 10129 Torino

D. Graffi, Via A. Murri 9, 40137 Bologna

A. Lagha-Benabdallah, Cité des annassers IV, Bat. 3 - app. n.8, Kouba, Alger

S. Larsson, Department of Mathematics, Chalmers University of Technology, S-412 96 Goteborg

A. Majda, Department of Mathematics, University of California, Berkeley, Cal. 94720

P. Marcati, Dipartimento di Matematica, Università, 38050 Povo (Trento)

C. Marchionna,, Viale Abruzzi 44, 20131 Milano

C. Marchioro, Dipartimento di Matematica, Università, 38050 Povo (Trento)

G. Matarazzo, Via Morelli e Silvati 62, 83100 Avellino

A.Maugeri, Via Etnea 688, 95128 Catania

A. Morro, Istituto Matematico Università, Via L.B. Alberti 4, 16132 Genova

G. Mulone, Via Sebastiano Catania 325, 95123 Catania

R. Nardini, Istituto Matematico Università, Piazza di Porta S. Donato 5, 40127 Bologna

M.C. Nucci, Istituto Matematico Università, Via Vanvitelli 1, 06100 Perugia

M. Padula, Via Vaiani 60, 80010 Quarto (Napoli)

P. Pucci, Istituto Matematico Università, Via Pascoli, 06100 Perugia

G. Raugel, 6 boulevard Jourdan, 75014 Paris

R. Salvi, Istituto Matematico del Politecnico, Via Bonardi 9, 20133 Milano

G. Salzano, I.A.C., Viale del Policlinico 137, 00161 Roma

P. Secchi, Dipartimento di Matematica, Università, 38050 Povo (Trento)

H. Sellers, 3003 N. Charles - Apt 2-G, Baltimore, MD 21218

J. Serrin, Univ. of Minnesota, School of Mathematics, 127 Vincent Hall, 206 Church Street S.E., Minneapolis, Minn. 55455

D. Socolescu, Institut f. ang. Math., Univ. Karlsruhe, Englerstr. 2, D 7500 Karlsruhe 1

Z. Tutek, Department of Mathematics, University of Zagreb, P.O.Box 187, Marulicev trg 19, 41001 Zagreb

A. Valli, Dipartimento di Matematica, Università, 38050 Povo (Trento)

N. Virgopia, Istituto Matematico Università, Città Universitaria, 00100 Roma

TABLE OF CONTENTS

C. BARDOS, *Introduction aux problèmes hyperboliques non linéaires* 1

A. MAJDA, *Smooth solutions for the equations of compressible and incompressible fluid flow* 75

G. GEYMONAT - P. LEYLAND, *The linear transport operator of fluid dynamics* ... 127

A. LAGHA-BENABDALLAH, *Limites des equations d'un fluide compressible lorsque la compressibilité tend vers zero* 139

C. MARCHIORO, *Vortex theory and Euler and Navier-Stokes evolution in two dimensions* 167

A. VALLI, *Free boundary problems for compressible viscous fluids* 175

INTRODUCTION AUX PROBLEMES HYPERBOLIQUES NON LINEAIRES

C. BARDOS

Département de Mathématiques - C.S.P. Université
de Paris-Nord, Av. J.B. Clément, 93430 Villetaneuse
et
Centre de Mathématiques Appliquées
Ecole Normale Supérieure
45, rue d'Ulm - 75005 Paris

INTRODUCTION

L'objet de cet exposé est de suivre au maximum les relations existant entre les notions d'entropies et les solutions faibles des systèmes hyperboliques. En évitant les détails techniques, on se propose de donner ainsi des résultats d'existence d'unicité et d'approximation. Le cas scalaire est traité complètement, selon les idées de Kruckov, car il est exemplaire et facile.

Au paragraphe II, on introduit la notion d'entropie pour les systèmes et on montre les relations entre cette notion et la notion de système symétrisable.

Le paragraphe III est essentiellement descriptif, on y regarde les propriétés du p système qui est à la fois assez simple pour donner lieu à plusieurs résultats explicites, et qui contient cependant beaucoup de phénomènes généraux.

Le paragraphe IV est enfin consacré à l'introduction des entropies généralisées et à leur utilisation pour prouver la convergence des solutions d'équations avec viscosité, selon les idées de Di Perna.

La rédaction de ce texte a suivi le cours CIME, il a donc été influencé par les commentaires et les réactions des autres participants.

Je tiens à remercier en particulier, pour leur aide et leur intérêt, A. Majda, J. Serrin, P. Geymonat et H. Da Veiga à qui, en plus du travail scientifique, revinrent les tâches d'organisation matérielle, qui rendirent mon séjour au CIME 82 particulièrement agréable.

TABLE DES MATIERES

INTRODUCTION GENERALE

I.- LOI DE CONSERVATION SCALAIRE 3

I.1.- Introduction 3

I.2.- Caractéristiques, vitesse finie de propagation et apparition de singularités 3

I.3.- Solution faible, condition de choc et d'entropie 5

I.4.- Théorèmes d'existence et d'unicité de la solution entropique pour une loi de conservation scalaire 10

Commentaires sur le paragraphe I. 16

II.- SOLUTIONS REGULIERES POUR LES SYSTEMES HYPERBOLIQUES A PLUSIEURS INCONNUES ET INTRODUCTION DE L'ENTROPIE 19

II.1.- Introduction 19

II.2.- Existence et unicité de la solution d'un système symétrisable 19

II.3.- Introduction de la notion d'entropie pour les systèmes 24

Commentaires sur le paragraphe II. 31

III.- UN EXEMPLE DE SYSTEME 2 × 2 A UNE DIMENSION D'ESPACE, LE p SYSTEME 33

III.1.- Introduction 33

III.2.- Généralités sur le p système 33

III.3.- Exemples de solutions élémentaires du problème de Riemann 38

III.4.- La méthode de Glimm 40

III.5.- Un théorème d'unicité pour les solutions faibles du p système 45

Commentaires sur le paragraphe III. 50

IV.- LES ENTROPIES APPROCHEES ET LA CONVERGENCE VERS UNE SOLUTION FAIBLE PAR COMPACITE PAR COMPENSATION 56

IV.1.- Introduction 56

IV.2.- Construction d'entropies approchées 56

IV.3.- Application de la notion d'entropie approchée à la convergence 58
faible des solutions de l'équation avec diffusion 58

IV.4.- Les majorations a priori uniformes. 65

Commentaires sur le paragraphe IV. 68

BIBLIOGRAPHIE 72

I.- LOI DE CONSERVATION SCALAIRE

1.- INTRODUCTION

Il est agréable de commencer un cours sur les problèmes hyperboliques par le cas des équations scalaires, car l'algèbre est particulièrement simple et permet une description explicite des phénomènes liés à l'hyperbolicité.

Les résultats concernant l'existence et l'unicité de la solution peuvent être facilement obtenus en particulier à l'aide du principe du maximum et de la multiplication par la fonction signe. Il en résulte que les méthodes utilisées ne peuvent pas, en général, être appliquées aux systèmes, sauf pour le cas très particulier des équations de Hamilton-Jacobi. Néanmoins, les lois scalaires permettent d'exhiber un certain nombre de pathologies fondamentales dues à la non linéarité.

2.- CARACTERISTIQUES, VITESSE FINIE DE PROPOGATION ET APPARITION DES SINGULARITES

On se propose d'étudier le problème de Cauchy pour des équations de la forme

$$\frac{\partial u}{\partial t} + \sum_{i=1}^{n} a_i(u) \frac{\partial u}{\partial x_i} = 0 \quad . \tag{I.1}$$

Les fonctions $u \to a_i(u)$ sont supposées assez régulières et on notera a le champ de vecteur de composantes $(a_i(u))_{1 \leq i \leq n}$. La donnée initiale $u(x,o)$ sera notée $\phi(x)$, elle sera de régularité variable selon les exemples. (I.1) est une équation scalaire dont la solution $u(x,t)$ est une fonction dépendant de n variables d'espace et d'une variable t le temps. A une variable d'espace (I.1) s'écrit aussi $\frac{\partial u}{\partial t} + f(u) \frac{\partial u}{\partial x} = 0$ et un cas particulier fondamental est fourni par l'équation de Burger $\frac{\partial u}{\partial t} + u \frac{\partial u}{\partial x} = 0$.

L'équation de Burger représente le mouvement idéal d'un fluide unidimensionnel en l'absence de forces extérieures, $u(x,t)$ désigne la vitesse de la particule située au point x à l'instant t, la trajectoire est donc donnée par l'équation différentielle

$$\dot{x}(t) = u(x(t),t) \quad . \tag{I.2}$$

Plus généralement, on appellera caractéristiques les solutions des équations différentielles :

(I.3) $\dot{x}(t) = f(u(x(t),t))$.

La relation (I.1) exprime que la fonction $t \to u(x(t),t)$ est constante. Il en est donc de même du second membre de (I.3).

Il en résulte que les caractéristiques sont des droites d'équation :

(I.4) $x(t) = x_o + t\, f(\phi(x_o))$.

La résolution du problème de Cauchy pour t assez petit est alors immédiate. Cette résolution a une interprétation géométrique simple, que nous explicitons sur l'exemple de l'équation de Burger.

On introduit la surface réglée Σ définie dans $\mathbb{R}_x \times \mathbb{R}_t \times \mathbb{R}_z$ par

(I.5) $\Sigma = \{(x,t,z) = (x, t\phi(x_o), t \, . \, \phi(x_o))$.

Pour $|t|$ assez petit, la projection de Σ sur le plan (x,t) est univoque et la solution est donnée par $u(x,t) = z$ où z est défini par la relation $(x,t,z) \in \Sigma$. (cf. figure I.1).

On peut alors déduire de la construction ci-dessus deux conséquences importantes.

1.- Les solutions se propagent à vitesse finie : plus précisément si ϕ est une fonction continûment différentiable sur $\mathbb{R}$, uniformément bornée ($|\phi(x)| \leq M$, $\forall\, x \in \mathbb{R}^n$) la pente des caractéristiques est majorée en valeur absolue par le nombre $N = \sup_{|\xi| \leq M} |f(\xi)|$. Il en résulte que la valeur de la solution au point (x,t) ne dépend que de la valeur de la donnée initiale dans la boule $B = \{X | \; |X - x| \leq Mt\}$.

En particulier, si ϕ est nulle en dehors de la boule $|x| \leq R$, $u(x,t)$ est nulle en dehors du cône $C = \{(x,t) | \; |x| \leq R + M|t| \}$.

En dimension 1, on peut également considérer une donnée initiale ϕ égale à la constante ϕ_- pour $x < -R$,et égale à la constante ϕ_+ pour $x > R$. Un raisonnement analogue montre alors que $u(x,t)$ sera égal à ϕ_- dans la région $\Sigma_- = \{(x,t) \,|\, x < -R - M|t|\}$ et à ϕ_+ dans la région $\Sigma_+ = \{(x,t) \,|\, x > R + M|t|\}$ (cf. figure I.2).

A la vitesse de propagation et à la non-linéarité est associé un phénomène nouveau qui est l'apparition de singularités. En effet, en dimension un d'espace l'existence de deux points x_- et x_+ vérifiant la relation $f(\phi(x_-)) > f(\phi(x_+))$

entraîne la collision des caractéristiques $X_- = x_- + tf(\phi(x_-))$ et $X_+ = x_+ + tf(\phi(x_+))$ et la solution ne peut plus être déterminée par la méthode précédente. Au point de rencontre elle devrait choisir entre les valeurs $f(\phi(x_+))$ et $f(\phi(x_-))$!

Une autre manière de montrer l'apparition de singularités consiste à prouver que la dérivée de la solution ne peut pas rester bornée dans $L^\infty(\mathbb{R}_x)$.

Supposons, pour simplifier, que f soit une fonction uniformément croissante (i.e. il existe une constante $\alpha > 0$ telle que l'ont ait $f'(\xi) > \alpha \quad \forall \xi \in \mathbb{R}$). En dérivant par rapport à x, on obtient pour la fonction $v(t) = -\frac{\partial u}{\partial x}(x(t),t)$ la relation

$$\text{(I.6)} \qquad \frac{dv}{dt} = f'(u(x(t),t))\left(\frac{\partial u}{\partial x}(x(t),t)\right)^2 \geq \alpha v^2 .$$

Il en résulte que v deviendra infinie au bout d'un temps fini dès qu'il existera un point ξ tel que l'on ait :

$$\text{(I.7)} \qquad \frac{\partial u}{\partial x}(\xi,0) = \phi'(\xi) < 0.$$

Ceci correspond bien à la situation décrite ci-dessus (i.e. l'existence de deux points x_- et x_+ vérifiant $f(\phi(x_-)) > f(\phi(x_+))$).

L'apparition de chocs conduit donc à introduire de nouvelles classes de solutions. Il s'agira bien sûr de solutions au sens des distributions. Il convient de souligner sur ces exemples l'importance de la notion de solutions faibles pour les problèmes non linéaires.

I .3.- SOLUTION FAIBLE, CONDITION DE CHOC ET D'ENTROPIE

Désormais, on se limitera à considérer le problème de Cauchy pour des temps positifs. On introduit des primitives $F(\xi)$ des fonctions $f_i(\xi)$ et on notera F le champ $F(\xi) = (f_1(\xi), f_2(\xi), \ldots, f_n(\xi))$.

On dira que u est une solution faible du problème de Cauchy si elle vérifie la relation :

$$\text{(I.8)} \qquad \frac{\partial u}{\partial t} + \sum_{i=1}^{n} \frac{\partial}{\partial x_i}(F_i(u)) = 0, \; u(x,0) = \phi(x).$$

Remarque 1.- Il sera désormais commode d'utiliser la notation $\nabla . F(u)$ pour la fonction $\sum_{i=1}^{n} \frac{\partial}{\partial x_i}(F_i(u))$.

Remarque 2.- Compte tenu de l'équation (I.8), $\frac{\partial u}{\partial t}$ est définie dans un espace de distribution convenable, il en résulte que la fonction $t \to u(.,t)$ est continue, à valeur dans ce même espace et donc que la relation $u(.,0) = \phi(.)$ a bien un sens.

Supposons que u soit une fonction continûment différentiable en dehors d'une surface orientable Σ de l'espace $\mathbb{R}^n_x \times \mathbb{R}^+_t$. Désignons par ν la normale à Σ et par u^+ et u^- les limites de u de part et d'autre de Σ. Alors l'équation (I.8) est équivalente aux deux assertions suivantes

(i) en dehors de Σ on a, au sens usuel, l'équation :

$$\text{(I.9)} \qquad \frac{\partial u}{\partial t} + \sum_{i=1}^{n} f_i(u) \frac{\partial u}{\partial x_i} = 0 .$$

(ii) sur Σ on a la formule des "sauts" :

$$\text{(I.10)} \qquad \nu_t(u^+ - u^-) + \nu_x(F(u^+) - F(u^-)) = 0.$$

(Dans la formule (I.10) ν_t et ν_x désignent respectivement les composantes de la normale ν selon l'axe des temps et selon le plan $\mathbb{R}^n_x$).

La relation (I.10) s'appelle relation de Rankine-Hugoniot.

Remarque 3.- Dans le cas d'une dimension d'espace, on convient d'orienter ν_x de la gauche vers la droite et la relation (I.10) s'écrit alors :

$$\text{(I.11)} \qquad F(u^+) - F(u^-) = -(\nu_x/\nu_t)(u^+ - u^-) ,$$

ou, en désignant par x(t) l'équation de la courbe Σ :

$$\text{(I.12)} \qquad \dot{x}(t) = (F(u^+) - F(u^-))/(u^+ - u^-) .$$ (1)

Pour l'équation de Burger , on a $f(u) = u$ et $F(u) = u^2/2$. L'équation (I.12) s'écrit alors :

$$\text{(I.13)} \qquad \dot{x}(t) = \frac{1}{2}(u^+ + u^-).$$

----

(1) F est une primitive de f.

On trouve que la vitesse de propagation du choc n'est autre que la moyenne des vitesses avant et après le choc.

L'analyse des conditions de Rankine-Hugoniot pour l'équation de Burger permet de fournir des contre-exemples à l'unicité. On choisit comme donnée initiale la fonction $\phi(x) = -\text{ sign } x$, alors pour tout nombre $a > 1$, les droites $x = (\frac{a-1}{2})t$ et $x = (\frac{1-a}{2})t$ sont respectivement de pentes positives et négatives et la fonction u définie par les relations

$$u(x,t) = \begin{cases} -1 & \text{si} \quad x > (\frac{a-1}{2})t \\ a & \text{si} \quad 0 > x > (\frac{a-1}{2})t \\ -a & \text{si} \quad 0 < x < (\frac{1-a}{2})t \\ +1 & \text{si} \quad x < (\frac{1-a}{2})t \end{cases}$$

est constante en dehors des droites de discontinuité et vérifie la relation de Rankine-Hugoniot sur les droites de discontinuité, c'est une solution de l'équation de Burger. Cette solution n'est pas physique, elle correspond à une indétermination liée au fait suivant : des caractéristiques sortent des lignes de choc (cf. figure I.3). La donnée initiale $\phi(x) = \text{sign } x$ représente un fluide dont toutes les particules situées à droite de zéro, sont animées de la vitesse -1, tandis que toutes les particules situées à gauche de zéro sont animées de la vitesse $+1$. La solution naturelle pour un tel fluide est l'équilibre :

$$u(t,x) = -1 \text{ si } x > 0, \quad u(t,x) = 1 \text{ si } x < 0,$$

qui est bien solution, au sens des distributions, de l'équation $\frac{\partial u}{\partial t} + \frac{\partial}{\partial x}(u^2/2) = 0$.

On se propose donc de préciser un critère d'unicité pour les solutions faibles exprimant que les caractéristiques rentrent dans le choc, et pour cela, on pose les définitions suivantes :

Définition 1.- A toute fonction régulière η on associe la fonction vectorielle q, flux d'entropie de la fonction η pour l'équation :

$$\text{(I.11)} \qquad \frac{\partial u}{\partial t} + \nabla \cdot F(u) = 0$$

en posant :

$$q_i^\eta(u) = \int_{\alpha_i}^{u} F_i'(s)\ \eta'(s)\ ds = \int_{\alpha_i}^{u} f_i(s)\ \eta'(s)\ ds\ ,$$

et on dira qu'une fonction est une solution entropique de l'équation si, pour toute fonction convexe η,elle vérifie, au sens des distributions, la relation

$$\text{(I.12)} \qquad \frac{\partial}{\partial t}\ \eta(u) + \nabla\ .\ q(u) \leq 0\ .$$

Remarque 4.- En multipliant par $\eta'(u)$ l'équation

$$\text{(I.13)} \qquad \frac{\partial u}{\partial t} + \sum_{i=1}^{n} a_i(u)\ \frac{\partial u}{\partial x_i} = 0$$

on obtient l'équation

$$\text{(I.14)} \qquad \frac{\partial}{\partial t}\ \eta(u) + \nabla\ .\ q(u) = 0$$

mais ce calcul n'est valable que pour des solutions régulières. Lorsqu'il apparaît des discontinuités les conditions de Rankine-Hugoniot sont modifiées par cette opération : par exemple les équations $\frac{\partial}{\partial t}\ u + \frac{\partial}{\partial x}\ \frac{u^2}{2} = 0$ et $\frac{\partial}{\partial t}\ (\frac{u^2}{2}) + \frac{\partial}{\partial x}\ (\frac{u^3}{3}) = 0$ ne donnent pas, le long des chocs, les mêmes conditions de Rankine-Hugoniot. Pour relier la notion d'entropie à la propriété des caractéristiques de rentrer dans le choc, nous prouvons la

Proposition 1.- Soit u une solution, au sens des distributions de l'équation

$$\text{(I.15)} \qquad \frac{\partial u}{\partial t} + \frac{\partial}{\partial x}\ F(u) = 0,$$

dans laquelle F désigne une fonction strictement convexe. On suppose que u est régulière en dehors d'une courbe de choc Σ donnée par la relation

$$\Sigma = \{(x,t)\ |\ x = x(t)\}$$

(l'énoncé et la démonstration sont identiques s'il s'agit d'un nombre fini de chocs)

alors les assertions suivantes sont équivalentes :

(i) pour toute fonction η, u vérifie la relation :

(I.16) $\quad \frac{\partial}{\partial t}\eta(u) + \frac{\partial}{\partial x} q^{\eta}(u) \leq 0$.

(ii) <u>il existe une fonction strictement convexe pour laquelle</u> u <u>vérifie la relation</u> (I.16).

(iii) <u>entre les caractéristiques et la courbe de choc</u>, <u>on a la relation</u> :

(I.17) $\quad f(u_i^-) \leq x_i'(t) \leq f(u_i^+)$.

(ce qui signifie que les caractéristiques rentrent dans le choc).

<u>Démonstration</u>.- Comme u est régulière en dehors de Σ (I.16) est équivalent à la condition de saut :

(I.18) $\quad \nu_t(\eta(u_+) - \eta(u_-)) + \nu_x(q^{\eta}(u_+) - q^{\eta}(u_-)) \leq 0$.

Soit, en utilisant la relation de Rankine-Hugoniot et la définition de q^{η} :

(I.19) $\quad E(u^+) = \left(\frac{F(u_+) - F(u_-)}{u_+ - u_-}\right)(\eta(u_+) - \eta(u_-)) - (q^{\eta}(u_+) - q^{\eta}(u_-)) \geq 0$.

Pour prouver la proposition, il suffit de montrer que si η est convexe, $E(u^+)$ est positif dès que l'on a $u^+ < u^-$ et réciproquement, que si η est strictement convexe, $E(u^+)$ est positif seulement si u^- est supérieur à u^+. Il suffit donc de montrer que la fonction $\xi \to E(\xi)$ qui vérifie $E(u^-) = 0$ est décroissante (strictement décroissante si η est strictement convexe). On a :

(I.20)
$$E'(\xi) = \frac{(F'(\xi)(\xi - u_-) - (F(\xi) - F(u_-)))}{(\xi - u_-)^2}(\eta(\xi) - \eta(u_-)) +$$
$$+ \frac{(F(\xi) - F(u_-))}{(\xi - u_-)}\eta'(\xi) - F'(\xi)\eta'(\xi)$$
$$= \frac{1}{(\xi - u_-)^2}[F(u_-) - F(\xi) - (u_- - \xi) F'(\xi)] (\eta(\xi) - \eta(u_-))] +$$
$$+ \frac{1}{(u_- - \xi)}[F(u_-) - F(\xi) - (u_- - \xi) F'(\xi)] (\eta'(\xi))$$
$$= \frac{1}{(\xi - u_-)^2}[F(u_-) - F(\xi) - (u_- - \xi) F'(\xi)][\eta(u_-) - \eta(\xi) - (u_- - \xi)\eta'(\xi)]$$

et la convexité de F et de η permet de conclure (figure I.3).

4.- THEOREME D'EXISTENCE ET D'UNICITE DE LA SOLUTION ENTROPIQUE POUR UNE LOI DE CONSERVATION SCALAIRE

Le fait que l'équation soit scalaire simplifie considérablement les démonstrations d'existence et d'unicité. Nous ne donnerons que les grandes lignes : le point essentiel est que l'espace adapté à décrire des solutions présentant des chocs est l'espace des fonctions à variation bornée, et que la solution peut être obtenue comme la limite, lorsque ε tend vers zéro, de la solution d'une équation parabolique avec un terme en $\varepsilon\Delta$.

Théorème 1.- Soit u_1 et u_2 deux fonctions mesurables, bornées sur $\mathbb{R}^\eta_x \times \mathbb{R}^+_t$ et à variation bornée par rapport aux variables x_i et t. On suppose que u_1 et u_2 sont deux solutions entropiques de l'équation

$$\frac{\partial u}{\partial t} + \nabla . F(u) = 0 \quad .$$

Alors il existe une constante M ne dépendant que de u_1, u_2 et de la fonction F telle que l'on ait :

$$\text{(I.21)} \qquad \int_{|x|\le R} |u_1(x,T) - u_2(x,T)|\,dx \le \int_{|x|\le R+MT} |u_1(x,0) - u_2(x,0)|\,dx.$$

Démonstration.- Pour toute fonction convexe η on a :

$$\text{(I.22)} \qquad \frac{\partial}{\partial t}\eta(u_i) + \nabla . q^\eta(u_i) \le 0 \qquad (i = 1,2) \quad .$$

En multipliant par des fonctions test, on constate que la régularité de η n'intervient pas, on peut donc faire tendre $\eta(\xi)$ vers la fonction convexe $|\xi - k_i|$ où k_i est une constante arbitraire. Le flux d'entropie correspondant est alors la fonction $(F(u) - F(k_i))$ sign $(u - k_i)$. On obtient alors les deux inéquations :

$$\text{(I.23)} \qquad \frac{\partial}{\partial t}|u_1 - k_1| + \nabla_x(F(u_1) - F(k_1)) \quad \mathrm{sgn}\ (u_1 - k_1)) \le 0$$

$$\text{(I.24)} \qquad \frac{\partial}{\partial \tau}|u_2 - k_2| + \nabla_y(F(u_2) - F(k_2)) \quad \mathrm{sgn}\ (u_2 - k_2)) \le 0$$

Dans la seconde on a changé le nom des variables (t,x) en les variables τ et y. Ceci permet de remplacer dans (I.23) k_1 par $u_2(\tau,y)$ et dans (I.24) k_2 par $u_1(t,x)$

ces arguments ne figurant que comme paramètres dans les équations correspondantes. On introduit ensuite une suite de fonctions positives δ_ε convergeant vers la masse de Dirac. Soit $\phi \in \mathcal{D}_+(\mathbb{R}^n \times \mathbb{R}_t^+)$, on applique le premier membre de (I.23) à la fonction

$$(x,t) \to \phi(\frac{x+y}{2}, \frac{t+\tau}{2})\, \delta_\varepsilon(x-y, t-\tau)$$

y,τ ne figurent alors que comme des paramètres. On intègre ensuite par rapport à ces paramètres. On applique ensuite le premier membre de (I.24) à la fonction

$$(y,\tau) \to \phi(\frac{x+y}{2}, \frac{t+\tau}{2})\, \delta_\varepsilon(x-y, t-\tau)$$

à nouveau dans cette opération ce sont les variables y et τ qui figurent comme des paramètres. On intègre enfin par rapport à ces variables et on obtient :

$$\text{(I.25)} \quad \iiiint (|u_1 - u_2| (\frac{\partial\phi}{\partial t} + \frac{\partial\phi}{\partial\tau}) + (F(u_1) - F(u_2)) \quad \mathrm{sgn}\,(u_1 - u_2)$$
$$. \, (\nabla_x\phi + \nabla_y\phi)\, \delta_\varepsilon(x-y, t-\tau)\, dx\, dy\, dt\, d\tau \geq 0.$$

En faisant tendre ε vers zéro, on force les variables x et y d'une part, et t et τ d'autre part à coïncider . On obtient ainsi la relation :

$$\text{(I.26)} \quad \frac{\partial}{\partial t}|u_1 - u_2| + \nabla_x((F(u_1) - F(u_2))\ \mathrm{sgn}\,(u_1 - u_2)) \leq 0 \quad .$$

On intègre enfin cette relation sur le tronc de cône

$$\Sigma = \{(x,t) \mid 0 \leq t \leq T, \quad |x| \leq R + M(T-t)\}$$

et il vient (au moins formellement ; le détail des calculs qui permet de justifier ces manipulations est dû à Kurckov [19]).:

$$\text{(I.27)} \int_{|x|\leq R} |u_1(x,T) - u_2(x,T)|\, dx$$
$$+ \int_{\delta\Sigma} \nu_t\, |u_1 - u_2| + \nu_x(F(u_1) - F(u_2))\ \mathrm{sgn}\,(u_1 - u_2) d\sigma \leq$$
$$\int_{|x|\leq R+MT} |u_1(x,0) - u_2(x,0)|\, dx \quad .$$

Dans le premièr membre de (I.27) $\partial\Sigma$ désigne la frontière latérale du tronc de cône Σ, ν_t et ν_x sont les composantes selon l'espace et le temps de la normale extérieure à $\partial\Sigma$ et ainsi on a :

$$\nu_t > 0 \quad \text{et} \quad \nu_t \geq M|\nu_x| \ .$$

Pour prouver la relation (I.21), il suffit maintenant de montrer que le second terme de (I.27) est positif.

Ceci s'obtient en écrivant l'inégalité :

$$\text{(I.28)} \quad \nu_t|u_1 - u_2| + \nu_x(F(u_1) - F(u_2)) \ \text{sgn}\ (u_1 - u_2)$$

$$\geq |u_1 - u_2|(\nu_t - |\nu_x| \sup_{|\xi| \leq (|u_1|_\infty, |u_2|_x)} \|(\nabla\ F)(\xi)\|)$$

$$\geq |u_1 - u_2|(\nu_t - |\nu_x|M) \geq 0.$$

(pourvu que M soit assez grand). □

Remarque 5.- Le théorème 1 implique bien entendu l'unicité des solutions entropiques. En fait, il est un peu plus précis car il contient un résultat de vitesse finie de propagation : ce qui se passe à l'instant T dans la boule de rayon R ne dépend que de ce qui se passe à l'instant zéro dans la boule de rayon R + MT. Mais bien entendu, la constante M dépend de la solution (en particulier de sa norme dans $L^\infty(\mathbb{R}^n_x \times \mathbb{R}_t)$).

Théorème 2.- On suppose que la donnée initiale ϕ appartient à $L^\infty(\mathbb{R}^n)$ et vérifie les relations :

$$\int_{\mathbb{R}^n} \left|\frac{\partial\phi}{\partial x_i}(x)\right| dx < +\infty \qquad \forall\, i \quad 1 \leq i \leq n$$

alors la solution u_ε de l'équation parabolique perturbée

$$\text{(I.28)} \quad \frac{\partial u_\varepsilon}{\partial t} + \nabla \,.\, A(u_\varepsilon) = \varepsilon\, \Delta u_\varepsilon, \quad u_\varepsilon(.,0) = \phi(.)$$

converge vers la solution entropique de l'équation

$$\frac{\partial u}{\partial t} + \nabla \,.\, A(u) = 0, \qquad u(.,0) = \phi(.) \ .$$

Remarque 6.- Comme on sait que la solution entropique, si elle existe est unique, il suffit de prouver qu'une sous-suite u_ε, extraite de la famille u_ε converge vers une solution entropique. D'autre part, l'équation (I.28) est parabolique non linéaire et il est facile de voir que cette équation admet une solution définie pour tout temps positif (Ladyzenskaia, Solonnikov et Uraltceva [18]), ainsi le théorème 2 permet de prouver l'existence d'une solution :

Démonstration du théorème 2

On remarque d'abord que, d'après le principe du maximum, pour tout $\varepsilon > 0$, on a

$$(I.30) \qquad |u_\varepsilon(x,t)| \leq |\phi(.)|_{L^\infty(\mathbb{R}^n_x)} \qquad \forall\, x,t\ .$$

On peut donc extraire de la famille u_ε une sous-famille encore notée u_ε convergeant vers une fonction $u \in L^\infty(\mathbb{R}^n_x \times \mathbb{R}^+_t)$, au sens suivant : pour tout compact K de $\mathbb{R}^n_x \times \mathbb{R}^+_t$ $u_{\varepsilon|K}$ converge vers $u_{|K}$ dans $L^\infty(K)$ faible $*$.

Le problème est alors de passer à la limite dans (I.28). Ceci est facile pour les termes linéaires $\partial u_\varepsilon/\partial t$ et $\varepsilon\,\Delta\, u_\varepsilon$, qui, au sens des distributions convergent respectivement vers $\partial u/\partial t$ et 0. Ceci est beaucoup plus difficile pour le terme non linéaire $\nabla\,.\,F(u\ )$.

En fait, il suffirait de prouver que $F(u_\varepsilon)$ converge dans $\mathcal{D}'(\mathbb{R}^n_x \times \mathbb{R}^+_t)$ vers $F(u)$, mais comme la fonction A est non linéaire, cela n'a rien d'évident.

Mais comme il s'agit d'un problème scalaire, on peut obtenir facilement des estimations a priori sur la variation totale, ce qui est l'objet du lemme 1. On terminera ensuite la démonstration du théorème en utilisant un argument de compacité.

Lemme 1.- La solution u_ε de l'équation parabolique perturbée :

$$\frac{\partial u_\varepsilon}{\partial t} + \nabla\ .\ F(u_\varepsilon) = \varepsilon\,\Delta\, u_\varepsilon, \qquad u_\varepsilon(.,0) = \phi(.)$$

satisfait les estimations a priori suivantes :

$$(I.31) \qquad \int_{\mathbb{R}^n} \left|\frac{\partial u_\varepsilon}{\partial x_i}(x,t)\right| dx \leq \int \left|\frac{\partial \phi}{\partial x_i}(x)\right| dx\ ,$$

$$\int_{\mathbb{R}^n} \left|\frac{\partial u_\varepsilon}{\partial t}(x,t)\right| dx \leq c \int_{\mathbb{R}^n} |\nabla\phi(x)|\,dx + \varepsilon \int_{\mathbb{R}^n} |\Delta\phi|\,dx$$

où la constante c ne dépend que de F et des données initiales.

Démonstration.- Encore une fois nous faisons une démonstration formelle, mais cette démonstration peut être rendue parfaitement rigoureuse en particulier en introduisant une suite de fonctions régulières $sgn_\eta(\xi)$ qui tendent, lorsque η

tend vers zéro vers la fonction sgn(ξ) définie par les relations :

$$\mathrm{sgn}(\xi) = \begin{cases} 1 & \text{si} \quad \xi > 0 \\ 0 & \text{si} \quad \xi = 0 \\ -1 & \text{si} \quad \xi < 0 \end{cases}$$

Ainsi on dérive l'équation perturbée, par rapport à toute variable x_i $(1 \le i \le n)$ et en posant $D_i u_\varepsilon = \frac{\partial u_\varepsilon}{\partial x_i}$, il vient :

$$\text{(I.32)} \qquad 0 = \int_{\mathbb{R}^n} \frac{\partial}{\partial t}\left(\frac{\partial u_\varepsilon}{\partial x_i}\right) \mathrm{sgn}\left(\frac{\partial u_\varepsilon}{\partial x_i}\right) dx + \int_{\mathbb{R}^n} \frac{\partial}{\partial x_i}(\nabla \cdot A(u_\varepsilon))\, \mathrm{sgn}\left(\frac{\partial u_\varepsilon}{\partial x_i}\right) dx$$

$$- \varepsilon \int_{\mathbb{R}^n} \frac{\partial}{\partial x_i}(\Delta u_\varepsilon)\, \mathrm{sgn}\left(\frac{\partial u_\varepsilon}{\partial x_i}\right) dx.$$

Le premier terme du second membre de (I.31) est égal à

$$\frac{d}{dt} \int_{\mathbb{R}^n} \left|\frac{\partial u_\varepsilon}{\partial x_i}\right| dx .$$

Pour le second membre, on intervertit les dérivées pour obtenir :

$$\text{(I.33)} \qquad \int_{\mathbb{R}^n} \frac{\partial}{\partial x_i} \frac{\partial}{\partial x_j} (A_j(u_\varepsilon))\, \mathrm{sgn}\left(\frac{\partial u_\varepsilon}{\partial x_i}\right) dx$$

$$= \int_{\mathbb{R}^n} \frac{\partial}{\partial x_j}\left(a_j(u_\varepsilon) \frac{\partial u_\varepsilon}{\partial x_i}\right) \mathrm{sgn}\left(\frac{\partial u_\varepsilon}{\partial x_i}\right) dx$$

$$= - \int_{\mathbb{R}^n} a_j(u_\varepsilon) \frac{\partial u_\varepsilon}{\partial x_i} \frac{\partial}{\partial x_j}\left(\mathrm{sgn}\left(\frac{\partial u_\varepsilon}{\partial x_i}\right)\right) dx.$$

et on remarque que le terme

$$\frac{\partial u_\varepsilon}{\partial x_i} \frac{\partial}{\partial x_j}\left(\mathrm{sgn}\left(\frac{\partial u_\varepsilon}{\partial x_i}\right)\right)$$

est "formellement" égal à zéro.

Enfin, le troisième terme est positif, car une intégration par partie donne :

$$(I.34) \quad -\varepsilon \int_{\mathbb{R}^n} \frac{\partial}{\partial x_i} \Delta u_\varepsilon \, \mathrm{sgn}\left(\frac{\partial u_\varepsilon}{\partial x_i}\right) dx = -\varepsilon \int_{\mathbb{R}^n} \frac{\partial}{\partial x_i} \Sigma \frac{\partial^2 u_\varepsilon}{\partial x_j^2} \mathrm{sgn}\left(\frac{\partial u_\varepsilon}{\partial x_i}\right) dx$$

$$= -\varepsilon \int_{\mathbb{R}^n} \Sigma \frac{\partial}{\partial x_j} \frac{\partial^2 u_\varepsilon}{\partial x_i \partial x_j} \mathrm{sgn}\left(\frac{\partial u_\varepsilon}{\partial x_i}\right) dx = \varepsilon \int_{\mathbb{R}^n} \mathrm{sgn}'\left(\frac{\partial u_\varepsilon}{\partial x_i}\right)\left(\frac{\partial^2 u_\varepsilon}{\partial x_i \partial x_j}\right)^2 dx,$$

et cette dernière expression est positive car $\mathrm{sgn}'\left(\frac{\partial u_\varepsilon}{\partial x_i}\right)$ est une distribution positive.

On a ainsi établi la relation :

$$\frac{d}{dt} \int_{\mathbb{R}^n} \left|\frac{\partial u_\varepsilon}{\partial x_i}(x,t)\right| dx \leq 0.$$

ce qui conduit à la première des inégalités de (I.31). Pour obtenir la seconde, il suffit de procéder de même, on dérive l'équation par rapport à t et on multiplie l'équation par sign $\left(\frac{\partial u_\varepsilon}{\partial t}\right)$; on obtient ainsi la relation :

$$\frac{d}{dt} \int_{\mathbb{R}^n} \left|\frac{\partial u_\varepsilon}{\partial t}\right| dx \leq 0,$$

d'où l'on déduit la seconde inégalité (I.31) en utilisant l'équation au temps $t = 0$. Si la fonction $\nabla \phi$ n'est pas intégrable, on remplace ϕ par $\phi_\varepsilon = \phi * \rho(\frac{\cdot}{\varepsilon}) \frac{1}{\varepsilon^n}$ où ρ est une fonction positive de masse totale 1, indéfiniment dérivable et à support compact. Toutes les majorations a priori précédentes restent valables et ϕ_ε converge vers ϕ tandis que $\varepsilon \int_{\mathbb{R}^n} |\Delta \phi_\varepsilon| \, dx$ reste borné indépendemment de ε.

<u>Fin de la démonstration du théorème 2</u>.- On va utiliser le lemme 1 pour montrer que l'on peut extraire de la suite u_ε une nouvelle sous-suite, encore notée u_ε convergeant vers u presque partout. Pour cela on construit une suite exhaustive de compacts de $\mathbb{R}^n_x \times \mathbb{R}^+_t$. Dans chacun de ces compacts, u_ε est borné dans $L^\infty(K)$ et dans $W^{1,1}(K)$, d'après le théorème de Kondrachov u_ε appartient donc à un compact de $L^\infty(K)$ et on peut en extraire une sous-suite qui converge presque partout. L'extraction d'une dernière suite, diagonale, conduit à la construction d'une famille u_ε bornée dans $L^\infty(\mathbb{R}^n_x \times \mathbb{R}^+_t)$ qui converge presque partout vers sa limite u. Le théorème de Lebesgue permet alors d'affirmer que u satisfait à l'équation :

$$\frac{\partial u}{\partial t} + \nabla \cdot F(u) = 0 .$$

Pour montrer que u est en fait une solution entropique, on utilise la convergeance presque partout et la régularisation elliptique. On a, pour toute fonction convexe η :

$$(I.35)\quad 0 = \eta'(u_\varepsilon)\left(\frac{\partial u_\varepsilon}{\partial t} + \nabla\ F(u_\varepsilon) - \varepsilon \Delta u_\varepsilon\right) = \frac{\partial}{\partial t}\eta(u_\varepsilon) + \nabla\ q^\eta(u_\varepsilon) - \varepsilon\,\Delta\,(\eta(u_\varepsilon)) + \varepsilon \sum_{i,j} \frac{\partial^2 \eta}{\partial x_i \partial x_j} \frac{\partial u_\varepsilon}{\partial x_i} \frac{\partial u_\varepsilon}{\partial x_j}\ .$$

On utilise alors la convexité de η pour obtenir la relation

$$(I.36)\quad \frac{\partial}{\partial t}\eta(u_\varepsilon) + \nabla \cdot q^\eta(u_\varepsilon) \leq 0.$$

Le théorème de Lebesgue permet alors de passer à la limite. On a ainsi montré que u est une solution entropique.

COMMENTAIRES SUR LE PARAGRAPHE 1

On a vu que bien que simples, les lois de conservations scalaires présentent une série de phénomènes importants.

La démonstration de l'apparition de singularités est très simple, néanmoins sous ses deux formes (rencontre de caractéristiques ou explosion de la dérivée première), elle sert de modèle aux démonstrations concernant l'apparition de singularités pour les systèmes ; on utilise alors les invariants de Riemann (introduits au § II) (cf. Lax [24] ou Klainerman et Majda [18]). L'apparition des singularités pour les équations scalaires à plus d'une dimension d'espace a été étudiée par D. Hopf [15].

Le théorème d'existence exposé ici est dû à Kruckov [21], il conduit à introduire l'espace BV des fonctions à variations bornées dont on trouvera une étude exhaustive dans Volpert [39]. Il s'agit de l'espace des fonctions mesurables et bornées dont les dérivées sont des mesures, ou ce qui revient au même l'espace des fonctions $u \in L^\infty(\mathbb{R}_x \times \mathbb{R}_t^+)$ limites, au sens des distributions de suites de fonctions u_ε vérifiant la majoration (uniforme par rapport à ε)

$$\int_{-\infty}^{\infty}\int_{-\infty}^{\infty} |\nabla u_\varepsilon(x,t)|\,dx\,dt + \sup_{x,t} |u_\varepsilon(x,t)| \leq C\ .$$

De même les théorèmes d'unicité et d'existence de solutions entropiques restent valables lorsqu'on suppose seulement que la donnée initiale est une fonction mesurable bornée et à variation bornée. Ceci permet de considérer en particulier des données initiales discontinues.

L'espace de fonctions à variation bornée convient bien à la description des solutions avec choc. En effet, pour toute fonction u à variation bornée sur un ouvert Ω il existe une partition de Ω en trois ensembles possédant les propriétés suivantes :

$$\Omega = \mathcal{C} \cup \mathcal{J} \cup \mathcal{R}$$

$\mathcal{J}$ est une réunion dénombrable de courbes continues, $\mathcal{R}$ est la réunion des extrémités de ces courbes car c'est un ensemble dénombrable et $\mathcal{J} \cup \mathcal{R}$ est un ensemble de mesure nulle et sur $\mathcal{C}$ complémentaire de $\mathcal{J} \cup \mathcal{R}$,u est une fonction continue. De plus en tout point de $\mathcal{J}$, u admet une limite à gauche et à droite.

On peut se demander si les solutions des lois de conservation qui sont continues dans l'ensemble $\mathcal{C}$ sont en fait, dans cet ensemble, plus régulières. Ceci n'est pas toujours vrai, mais est prouvé pour un ensemble de données initiales génériques par Golubitsky et Schaeffer [14].

Le cadre des lois scalaires permet d'analyser complètement les méthodes numériques (cf. Leroux [25,26]), c'est d'ailleurs à partir d'une méthode de différences finies qu'a été prouvée(avant Kruckov [16]) à plus d'une dimension d'espace, l'existence de solutions (cf. Conway et Smoller [3]).

A l'existence et à l'unicité des solutions, sont associées des propriétés de contraction non linéaire dans les espaces $L^1(\mathbb{R}^n)$, ce qui conduit à leur associer des semi-groupes non linéaires à contraction dans cet espace. C'est le premier exemple d'application du formalisme introduit par Crandall et Ligget [7] (cf. aussi Quinn-Keyfitz [36]).

A une variable d'espace les lois de conservation scalaires coïncident avec l'équation de Hamilton Jacobi $\frac{\partial \phi}{\partial t} + H\left(\frac{\partial \phi}{\partial x}\right) = 0$. En fait, la théorie de Kruckov s'étend de manière systématique aux équations de Hamilton-Jacobi :

$$\frac{\partial \phi}{\partial t} + H(\nabla \phi) + g(x) = 0 .$$

On pourra consulter pour cela le livre de P.L. Lions [27] et les références qui s'y trouvent.

Lorsque f est une fonction convexe, on peut obtenir une représentation explicite de la solution de $\frac{\partial u}{\partial t} + \frac{\partial}{\partial x} f(u) = 0$, à l'aide d'une minimisation, cela a été fait par Hopf [16] et Lax [23] et repris pour l'équation d'Hamilton-Jacobi dans un cadre analogue (cf. P..L. Lions [27]).

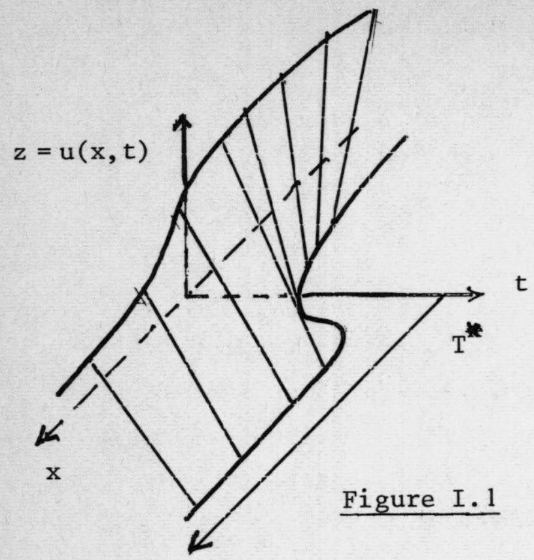

Figure I.1

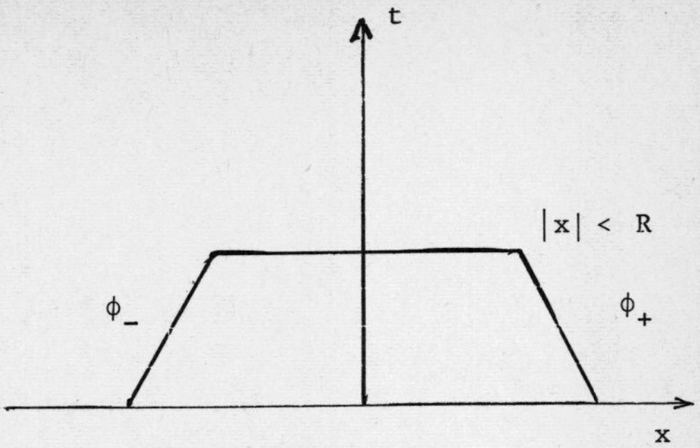

Figure I.2

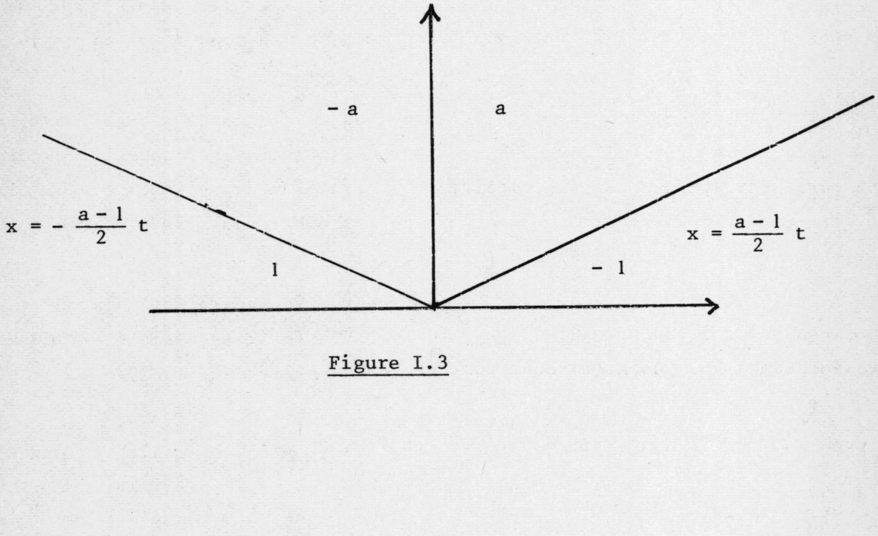

Figure I.3

II.- SOLUTIONS RÉGULIÈRES POUR LES SYSTÈMES HYPERBOLIQUES À PLUSIEURS INCONNUES ET INTRODUCTION DE L'ENTROPIE

II.1.- INTRODUCTION

On ne connaît que très peu de résultats pour les systèmes hyperboliques à plus d'une variable d'espace et à plus d'une inconnue, ceci essentiellement à cause de l'apparition de singularités au bout d'un temps fini (comme en dimension 1) et à la structure complexe que peut présenter l'ensemble de ces singularités (naturellement beaucoup plus complexe en dimension supérieure à 1). Dans la phase de régularité, la théorie des problèmes non linéaires présente des similarités avec celle des problèmes hyperboliques linéaires, et par des méthodes de perturbation, on peut prouver des théorèmes du type Cauchy-Kowalewsky ; mais même dans ce cadre, il est intéressant d'introduire la notion d'entropie d'autant que celle-ci permet de prouver un théorème d'unicité mettant en jeu à la fois une solution régulière et toute solution faible vérifiant la condition d'entropie.

II.2.- EXISTENCE ET UNICITE DE LA SOLUTION D'UN SYSTEME SYMETRISABLE

On se propose dans ce paragraphe, d'étudier le problème de Cauchy (existence et unicité locale) pour des systèmes de la forme :

$$\text{(II.1)} \qquad \frac{\partial}{\partial t} + \sum_{i=1}^{n} A_i(U) \frac{\partial U}{\partial x_i} = 0 \; .$$

La dimension de l'espace est en général égale à 1,2,3 et U désigne une fonction à valeur dans $\mathbb{R}^m$. Les $A_i(U)$ sont donc des applications d'un ouvert $\Omega \subset \mathbb{R}^m$ à valeur dans les matrices carrées réelles M_m (à m lignes et m colonnes).

On supposera que sur Ω les fonctions $U \to A_i(U)$ sont indéfiniment différentiables et on dira que le système est <u>symétrisable</u> s'il existe une matrice $A_o(U)$ possédant les propriétés suivantes ;

(1) $U \to A_o(U)$ est une application indéfiniment différentiable de Ω dans M_m ;

(2) Pour tout $U \in \Omega$, $A_o(U)$ est une matrice symétrique définie et positive ;

(3) Pour tout i, $1 \le i \le n$, les matrices $A_o(U)$, $A_i(U)$ sont symétriques.

Remarque 1.- Soit $\omega \in S^{n-1}$ et $\lambda_k(U,\omega)$ $(1 \leq k \leq m)$ les valeurs propres de la matrice $\sum_{i=1}^{n} A_i(U)\omega_i$. Comme $A_o(U)$ est inversible, on déduit de la relation :

$$(II.2) \qquad 0 = \det(\lambda I - \sum_{i=1}^{n} A_i(U)\omega_i) = \det(\lambda A_o(U) - \sum_{i=1}^{n} A_o(U)\, A_i(U)\omega_i)$$

que les valeurs propres $\lambda_k(U,\omega)$ sont toutes réelles , tout système symétrisable est (au sens usuel) hyperbolique ; mais comme on le verra, la notion de système symétrisable est bien mieux adaptée aux problèmes non linéaires. En particulier, on obtient des résultats simples d'unicité et d'existence décrits dans les théorèmes suivants :

Théorème 1. On désigne par U_1 et U_2 deux solutions définies sur $\mathbb{R}^n_x \times]-T,T[$ à valeurs dans Ω. On suppose que sur $\mathbb{R}^n_x \times]-T,T[$, U_1 et U_2 sont uniformément lipschitziennes et sont solutions au sens usuel du système symétrisable :

$$(II.3) \qquad \frac{\partial U}{\partial t} + \sum_{i=1}^{n} A_i(U) \frac{\partial U}{\partial x_i} = 0.$$

Il existe alors une constante $M > 0$ telle que si U_1 et U_2 coïncident pour $t = 0$ dans la boule $|X| \leq R$, elles coïncident dans le domaine de $\mathbb{R}^n_x \times \mathbb{R}_t$ défini par :

$$Q = \{(x,t) \mid x \in \mathbb{R}^n,\ |t| < T\ ;\ |x| \leq R - M|t|\}.$$

On a ainsi un résultat d'unicité et de vitesse finie de propagation.

Démonstration du théorème 1.- On fait la démonstration pour $t > 0$, la démonstration pour $t < 0$ est identique. Comme U_1 et U_2 sont des solutions, on a la relation :

$$(II.4) \qquad (A_o(U_1) \frac{\partial U_1}{\partial t} - A_o(U_2) \frac{\partial U_2}{\partial t}\ ,\ U_1 - U_2)$$

$$+ \sum_{i=1}^{n} (A_o(U_1)\, A_i(U_1) \frac{\partial U_1}{\partial x_i} - A_o(U_2)\, A_i(U_2) \frac{\partial U_2}{\partial x_i}\ ,\ U_1 - U_2) = 0,$$

où les (,) désignent le produit scalaire euclidien dans $\mathbb{R}^m$. On pose ensuite $B_i(U) = A_o(U)\, A_i(U)$, et on écrit :

(II.5) $$(A_o(U_1)\frac{\partial U_1}{\partial t} - A_o(U_2)\frac{\partial U_2}{\partial t}, U_1 - U_2) = ((A_o(U_1) - A_o(U_2))\frac{\partial U_1}{\partial t}, U_1 - U_2)$$

$$+ (A_o(U_2)\frac{\partial}{\partial t}(U_1 - U_2), U_1 - U_2) = \frac{1}{2}\frac{\partial}{\partial t}(A_o(U_2)(U_1 - U_2), U_1 - U_2)$$

$$+ ((A_o(U_1) - A_o(U_2))\frac{\partial U_1}{\partial t} - \frac{1}{2}A_o'(U_2)\,.\,\frac{\partial U_2}{\partial t}\,.\,U_1 - U_2), U_1 - U_2).$$

De même on a :

(II.6) $$(B_i(U_1)\frac{\partial U_1}{\partial x_i} - B_i(U_2)\frac{\partial U_2}{\partial x_i}, U_1 - U_2) = \frac{1}{2}\frac{\partial}{\partial x_i}(B_i(U_2)(U_1 - U_2), (U_1 - U_2))$$

$$((B_i(U_1) - B_i(U_2))\frac{\partial U_1}{\partial x_i} - \frac{1}{2}B_i'(U_2)\,.\,\frac{\partial U_2}{\partial x_i}\,.\,U_1 - U_2, U_1 - U_2).$$

De (II.5) et (II.6), on déduit la relation :

(II.7) $$\frac{\partial}{\partial t}(A_o(U_2)(U_1 - U_2), (U_1 - U_2)) + \sum_{i=1}^{n}\frac{\partial}{\partial x_i}(B_i(U_2)(U_1 - U_2), (U_1 - U_2))$$

$$= F(\nabla U_1, \nabla U_2, U_1 - U_2)$$

où $F(\nabla U_1, \nabla U_2, U_1 - U_2)$ vérifie, pour ∇U_1 et ∇U_2 uniformément bornés, une majoration du type :

(II.8) $$|F(\nabla U_1, \nabla U_2, U_1 - U_2)| \leq C|U_1 - U_2|^2 .$$

L'intégration de (II.7) sur le cône C_t défini par :

(II.9) $$C_t = \{(x,s) \mid 0 \leq s \leq t \mid |x| \leq R - M_s\}$$

donne alors la relation :

$$(\text{II.10})\quad \int_{|x|\leq R-MT} (A_o(U_2)(U_1-U_2),(U_1-U_2))(x,t)\,dx$$

$$+\int_o^t ds \int_{\substack{|x|=R-Ms\\ 0\leq s\leq t}} ((A_o(U_2)\nu_t+\Sigma\,\nu_i\;B_i(U_2))(U_1-U_2),(U_1-U_2))\,ds$$

$$\int_{|x|=R} (A_o(U_2)(U_1-U_2),(U_1-U_2))(x,0)\,dx + C\int_o^t \int_{|x|\leq R-Ms} \|U_1-U_2\|^2 dx\; ds.$$

Dans la seconde intégrale du premier membre de (II.10), ν_t et ν_i désignent les composantes extérieures de la normale au cône $|x| = R-Ms$. Comme A_o est définie positive, pour M assez grand, la matrice

$$\nu_t\;\; A_o(U_2) + \Sigma\;\nu_i\;\; B_i(U_2)$$

est définie positive. Enfin, on utilise le lemme de Gronwall et la relation

$$(A_o(U_2)\;U_1-U_2,U_1-U_2)\;\geq\;\gamma|U_1-U_2|^2$$

pour terminer la démonstration.

Le théorème d'existence utilise le même type de majorations a priori mais comme il faut contrôler la norme de Lipschitz de U par des intégrales de type énergie, on utilise le théorème de Sobolev en introduisant une majoration des dérivées d'ordre p où p désigne le plus petit entier supérieur à $\frac{n}{2}+1$.

Théorème 2.- On désigne par p le plus petit entier supérieur à $n/2+1$; on note $|||\cdot|||$ la norme dans l'espace de Sobolev $H^p(\mathbb{R}^n_x)$, et par D la constante d'injection de Sobolev de l'espace $H^p(\mathbb{R}^n_x)$ dans $L^\infty(\mathbb{R}^n_x)$. On suppose que Ω contient une boule de centre O et de rayon $R>0$ et que pour $U\in B(O,R)$ le système

$$(\text{II.11})\qquad \frac{\partial U}{\partial t}+\sum_{i=1}^{n} A_i(U)\,\frac{\partial U}{\partial x_i}=0$$

est symétrisable alors il existe une constante C telle que pour toute donnée initiale $U_o\in H^p(\mathbb{R}^n_x)$ vérifiant, $|||U_o||| < R/D$ et tout T vérifiant :

$$(\text{II.12})\qquad T<\frac{1}{C|||U_o|||}\left(1-\left(C\,\frac{|||U_o|||}{R}\right)^{p-2}\right)$$

le système (II.11) admette dans $\mathbb{R}^n_x\times\,]-T,T[$ une solution vérifiant la condition initiale $U(.,0)=U_o(.)$ et la majoration :

(II.13) $\qquad |||U(.,t)||| \leq C|||U_o|||/(1 - C\, t|||U_o|||^{p-2})^{1/p-2}$.

<u>Démonstration</u>.- Comme la norme de U_o dans $H^p(\mathbb{R}^n)$ est choisie assez petite; on déduit du théorème de Sobolev la relation :

(II.14) $\qquad \|U_o\|_{L^\infty(\mathbb{R}^n_x)} \leqslant D\, C_1 < R$.

Donc pour tout x, $U_o(x)$ appartient à B(0,R). On construit alors la suite $U^k(x,t)$ par l'algorithme suivant :

(II.15) $\qquad U^o(x,t) = U_o(x)$,

$$\frac{\partial U^{k+1}}{\partial t} + \sum_{i=1}^{n} A_i(U^k) \frac{\partial U^{k+1}}{\partial x_i} = 0, \qquad U^{k+1}(0) = U_o(x).$$

Pour tout k le système figurant dans (II.15) est linéaire en U^{k+1}. Il est symétrisable, pourvu que pour tout couple (x,t) ($|t| < T$), $U^k(x,t)$ appartienne à B(0,R). Cette condition est réalisée pour k = 0. On va simultanément montrer de proche en proche qu'elle est réalisée pour tout k et établir une majoration du type (II.13) pour U^{k+1}. On dérive m fois (II.15), m est un multi-entier quelconque de longueur inférieure ou égale à p. On obtient une expression de la forme :

(II.16) $$\frac{\partial}{\partial t} D^m U^{k+1} + \sum_{i=1}^{n} A_i(U^k) \frac{\partial}{\partial x_i} (D^m U^{k+1}) =$$

$$- \sum_{i=1}^{m} \Big(\sum_{0<\ell<m} \binom{m}{\ell} D^\ell (A_i(U^k))\, D^{m-\ell} \frac{\partial U^{k+1}}{\partial x_i}\Big) .$$

On multiplie ensuite scalairement par $A_o(U^k)\, D^m U^{k+1}$. On intègre sur $\mathbb{R}^n \times [0,t]$ ($t > 0$, le cas $t < 0$ se traite de manière analogue). On utilise la symétrie des matrices $A_o(U^k)\, A_i(U^k)$ et le fait que $A_o(U^k)$ est définie positive. On obtient en sommant par rapport à m :

(II.17) $$|||U^{k+1}(.,t)|||^2 \leq C|||U_o(.)|||^2 + \int_o^t \int_{\mathbb{R}^n} G(U^k, DU^k, \ldots, D^m U^k, D^m U^{k+1})\,dx\,ds.$$

Le second terme du second membre de (II.16) contient des expresssions de la forme

$$\frac{1}{2} \Big(\frac{\partial}{\partial x_i} (A_i(U^k))\, D^m U^{k+1}, D^m U^{k+1}\Big)$$

et des expressions dela forme

$$(D^{\ell}(A_i(U^k))D^{m-\ell}\frac{\partial U^{k+1}}{\partial x_i}, \quad D^m U^{k+1}) \quad |\ell| \geq 1.$$

En utilisant les théorèmes d'injection de Sobolev, on montre que ce second terme est majoré par une expression de la forme

$$C|||U^k|||^p \; |||U^{k+1}||| \; .$$

On obtient donc une inégalité du type Gronwall

$$\text{(II.18)} \qquad |||U^{k+1}(.,t)|||^2 \leq C|||U_o(.)|||^2 + C\int_o^t |||U^k(.,s)|||^p|||U^{k+1}(.s)||| \, ds$$

d'où l'on déduit la majoration a priori :

$$\text{(II.19)} \qquad |||U^{k+1}(.,t)||| \leq C|||U_o|||/(1-ct|||U_o|||^{p-2})^{1/p-2} \quad .$$

$U^{k+1}(x,t)$ appartiendra à Ω dès que l'on aura :

$$DC|||U_o|||/(1 - ct|||U_o|||^{p-2})^{-1/p-2} \leq R \quad .$$

Ceci est vérifié pour t = 0 et C_1 assez petit, et pour t positif, pourvu que l'on ait

$$\text{(II.20)} \qquad t \leq \frac{1}{C|||U_o|||}\left(1 - \left(\frac{C|||U_o|||}{R}\right)^{p-2}\right).$$

On peut ainsi réitérer le procédé en changeant k en k + 1 et prouver pour U^{k+2} une majoration du type (II.19). Il est enfin facile de montrer que la suite U^k converge vers une fonction U satisfaisant aux conclusions du théorème 1.

II.3.- INTRODUCTION DE LA NOTION D'ENTROPIE POUR LES SYSTEMES

On dit qu'un système est conservatif et hyperbolique s'il s'écrit sous la forme :

$$\text{(II.21)} \qquad \frac{\partial U}{\partial t} + \sum_{i=1}^{n} \frac{\partial}{\partial x_i}(F_i(U)) = 0$$

et si les valeurs propres de la matrice $\sum_{i=1}^{n} F_i'(U)\,{}_i$ sont pour tout U dans le domaine de définition du système, et pour tout $\omega \in S^{n-1}$, réelles.

On a ainsi un système de m équations avec n inconnues et on peut se demander si comme dans le cas scalaire il est possible d'exhiber des équations ou des inéquations supplémentaires liées à la notion d'entropie. On dira qu'une fonction $\eta(U)$ est une entropie si, pour toute fonction $F_i(U)$, il existe une fonction $G_i(U)$ telle que l'on ait

(II.22) $\eta'(U) \,.\, F_i'(U) = G_i'(U)$.

On déduit alors de l'équation (II.21) que toute solution régulière vérifie la relation

(II.23) $$\frac{\partial}{\partial t}\eta(U) + \sum_{i=1}^{n} \frac{\partial}{\partial x_i}(G_i(U)) = 0 \;;$$

les fonctions $G_i(U)$ sont appelées flux d'entropie. On a vu que dans le cas scalaire toute fonction était une entropie et que les entropies convexes jouaient un rôle fondamental dans la détermination de la solution faible. Il n'en est plus de même dans le cas vectoriel et à part les systèmes 2×2 (cf. § III et commentaires) les systèmes hyperboliques n'admettent en général pas d'entropie, sauf certains systèmes dérivés de la physique ; l'entropie ayant alors une signification physique. Ce sont ces exemples qui donnent un intérêt à la notion.

On note $F_i^k(U)$ la kième composante de la fonction $F_i(U)$; la relation (II.23) s'écrit alors sous la forme

(II.24) $$\frac{\partial G_i}{\partial u_\ell} = \sum_{1\le r\le n} \frac{\partial \eta}{\partial u_r}\frac{\partial F_i^r}{\partial u_\ell} \,.$$

Pour assurer l'existence de G il faut et il suffit que soit satisfaite la relation

(II.25) $$\frac{\partial}{\partial u_j}\Big(\sum_{1\le r\le m} \frac{\partial \eta}{\partial u_r}\frac{\partial F_i^r}{\partial u_\ell}\Big) = \frac{\partial}{\partial u_\ell}\Big(\sum_{1\le r\le m} \frac{\partial \eta}{\partial u_r}\frac{\partial F_i^r}{\partial u_j}\Big)$$

c'est-à-dire, compte tenu de l'égalité

$$\frac{\partial}{\partial u_j}\Big(\frac{\partial F_i^r}{\partial u_\ell}\Big) = \frac{\partial}{\partial u_\ell}\Big(\frac{\partial F_i^r}{\partial u_j}\Big)$$

la relation

(II.26) $$\sum_{1\le r\le m} \frac{\partial^2 \eta}{\partial u_r \partial u_j}\frac{\partial F_i^r}{\partial u_\ell} = \sum_{1\le r\le m} \frac{\partial^2 \eta}{\partial u_r \partial u_\ell}\frac{\partial F_i^r}{\partial u_j}$$

Le premier membre de (II.26) est le (ℓ,j) ième coefficient de la matrice $(F_i'(U))^* \eta''(U)$, tandis que le second membre de (II.26) est le (ℓ,j) ième coefficient de la matrice $\eta''(U)\, F_i'(U)$, la relation (II.26) s'écrit donc

(II.27) $\eta''(U)\, F_i'(U) = (F_i'(U))^* \eta''(U)$.

On a ainsi prouvé la

Proposition 1

(i) Pour que le système

(II.28) $$\frac{\partial U}{\partial t} + \sum_{i=1}^{n} \frac{\partial}{\partial x_i} (F_i(U)) = 0$$

admette pour entropie la fonction $\eta(U)$, il faut et il suffit que les matrices

$$\eta''(U)\, F_i'(U)$$

soient symétriques.

(ii) En particulier, tout système symétrique admet l'énergie $\eta(U) = \sum_{i=1}^{n} u_i^2$ comme entropie.

(iii) Inversement, tout système qui admet une entropie convexe est symétrisable il possède donc en particulier la propriété de vitesse finie de propagation.

Comme dans le cas scalaire, il est possible de relier l'entropie à la notion de viscosité limite par la

Proposition 2.- On suppose que le système $\frac{\partial U}{\partial t} + \Sigma \frac{\partial}{\partial x_i} (F_i(U))$ admet une entropie convexe $\eta(U)$. On désigne par U_ε $(\varepsilon > 0)$ la solution du système parabolique

(II.29) $$\frac{\partial U_\varepsilon}{\partial t} + \sum_{i=1}^{n} \frac{\partial}{\partial x_i} (F_i(U_\varepsilon)) = \varepsilon \Delta U_\varepsilon, \quad U_\varepsilon(x,0) = \Phi(x).$$

On suppose que lorsque ε tend vers zéro, U_ε reste borné dans $L^\infty(\mathbb{R}^n \times \mathbb{R}_t^+)$ et converge pour presque tout (x,t) vers une fonction U. Alors U est solution (au sens des distributions) du système)

(II.30) $$\frac{\partial U}{\partial t} + \sum_{i=1}^{n} \frac{\partial}{\partial x_i} (F_i(U)) = 0$$

et de l'inéquation d'entropie

(II.31) $$\frac{\partial}{\partial t} \eta(U) + \sum_{i=1}^{n} \frac{\partial}{\partial x_i} (G_i(U)) \leq 0.$$

Démonstration.- Compte tenu des hypothèses (existence d'une entropie convexe et surtout convergence presque partout de U) la démonstration de cette proposition est évidente. Comme U_ε converge presque partout et reste borné, $F_i(U_\varepsilon)$ converge vers $F_i(U)$ (dans $\mathcal{D}'$), ainsi $\frac{\partial}{\partial x_i}(F_i(U_\varepsilon))$ converge vers $\frac{\partial}{\partial x_i}(F_i(U))$. De même $\varepsilon \Delta U_\varepsilon$ converge vers zéro dans $\mathcal{D}'$. On en déduit que U est solution de (II.30). Ensuite, en multipliant par $\eta'(U_\varepsilon)$ l'équation (II.29) on obtient

$$\text{(II.31)} \quad \frac{\partial}{\partial t}\eta(U_\varepsilon) + \sum_{i=1}^{n} \frac{\partial}{\partial x_i}(G_i(U_\varepsilon)) = \varepsilon\, \eta''(U_\varepsilon)\, \Delta U_\varepsilon$$

$$= \varepsilon \Delta \eta(U_\varepsilon) - \varepsilon(\eta''(U_\varepsilon)\nabla U_\varepsilon\,,\ \nabla U_\varepsilon).$$

Soit

$$\text{(II.33)} \quad \frac{\partial}{\partial t}\eta(U_\varepsilon) + \sum_{i=1}^{n} \frac{\partial}{\partial x_i}(G_i(U_\varepsilon)) - \varepsilon \Delta \eta(U_\varepsilon) \le 0$$

et un nouveau passage à la limite donne (II.30).

Il est intéressant de remarquer enfin que la notion d'entropie contient un résultat facile d'unicité. Ce résultat est cependant faible car il compare une solution régulière et une solution éventuellement discontinue.

Théorème 3.- On considère le système

$$\text{(II.34)} \quad \frac{\partial U}{\partial t} + \sum_{i=1}^{n} \frac{\partial}{\partial x_i}(F_i(U)) = 0$$

On suppose que ce système admet une entropie strictement convexe $\eta(U)$, pour U à valeur dans un domaine Ω de $\mathbb{R}^n$.

On considère dans $\mathbb{R}^n_x \times [0,T]$ deux solutions U et V (à valeur dans Ω) possédant les propriétés suivantes :

(i) U est une solution régulière (i.e. une fonction dont les dérivées sont bornées)

(ii) V est une fonction mesurable et bornée, elle présente éventuellement des discontinuités et est donc solution de

$$\text{(II.35)} \quad \frac{\partial V}{\partial t} + \sum_{i=1}^{n} \frac{\partial}{\partial x_i} F_i(V) = 0.$$

On suppose de plus que V vérifie au sens des distributions l'inégalité d'entropie :

(II.36) $$\frac{\partial}{\partial t}\eta(V) + \sum_{i=1}^{n} \frac{\partial}{\partial x_i} G_i(V) \leq 0 .$$

Alors si U _et_ V _coïncident pour_ $t = 0$, _elle coïncident dans tout le domaine_ $\mathbb{R}^n_x \times [0,T^*[$. _Plus précisément_, _il existe une constante_ M _telle que pour tout_ $R > 0$ _et tout_ $0 < t < T^*$ _on ait_ :

(II.37) $$\int_{|x| \leq R-Mt} |U(x,t) - V(x,t)|^2 dx \leq \beta e^{\gamma t} \int_{|x|\leq R} |U(x,0) - V(x,0)|^2 dx.$$

_où_ β _et_ γ _désignent des constantes positives_.

Démonstration.- (II.37) énonce en fait un résultat de vitesse finie de propagation, aussi on introduira le tronc de cône C_t :

$$C_t = \{(x,t) \quad 0 \leq t \leq T < T^*, \quad |x| < R - Mt\} .$$

Comme U est une solution régulière, on a :

(II.38) $$\frac{\partial}{\partial t}\eta(U) + \sum_{i=1}^{n} \frac{\partial}{\partial x_i} G_i(U) = 0$$

et comme V satisfait la condition d'entropie, on a :

(II.39) $$\frac{\partial}{\partial t}\eta(V) + \sum_{i=1}^{n} \frac{\partial}{\partial x_i} G_i(V) \leq 0$$

ainsi on peut soustraire (II.38) de (II.39) pour obtenir :

(II.40) $$\frac{\partial}{\partial t}(\eta(V) - \eta(U) - \eta'(U)(V-U)) + \sum_{i=1}^{n} \frac{\partial}{\partial x_i}(G_i(V) - G_i(U)) \leq -\frac{\partial}{\partial t}(\eta'(U), V-U) \quad .$$

On va ensuite intégrer les deux membres de (II.40) sur le tronc ce cône C_t (ceci nécessite un peu de régularité sur V , mais l'hypothèse de régularité peut être facilement levée de la manière suivante : on introduit une famille C_ε de fonctions indéfiniment dérivables, positives, bornées, à support compact dans $\mathbb{R}^n_x \times \mathbb{R}^+_t$ et convergeant presque partout vers la fonction caractéristique de C_T. On applique à C_ε les deux membres de la distribution (II.40) en suivant les calculs qui vont être détaillés ci-dessous, puis on fait tendre ε vers zéro). On poursuit donc un calcul formel (ou nécessitant un peu de régularité sur V). On désignera toujours par Σ_T la frontière latérale du tronc de cône C_T ($\Sigma_T = (x,t) \mid 0 < t < T, |x| < R - Mt\}$) par $d\sigma$ l'élément d'axe sur Σ_T et par ν_t, ν_i

$(1 \le i \le n)$ les composantes de la normale extérieure à Σ_T dans $\mathbb{R}^n_x \times \mathbb{R}_t$. On a :

$$\text{(II.41)} \qquad \int_{C_T} \frac{\partial}{\partial x_i} (G_i(V) - G_i(U))\, dx\, dt = \int_{\Sigma_T} \nu_i (G_i(V) - G_i(U))\, d\sigma$$

et :

$$\text{(II.42)} \qquad \frac{\partial}{\partial t} (\eta'(U), V - U) = (\eta''(U) \frac{\partial U}{\partial t}, V - U) + (\eta'(U), \frac{\partial}{\partial t}(V - U)) =$$

$$= - (\eta''(U) \sum_{i=1}^{n} F_i'(U) \frac{\partial U}{\partial x_i}, V - U) - (\eta'(U), \sum_{i=1}^{n} \frac{\partial}{\partial x_i}(F_i(V) - F_i(U)).$$

On intègre alors le second terme du dernier membre de (II.42) sur C_t pour obtenir :

$$\text{(II.43)} \qquad - \int_{C_T} (\eta'(U), \sum_{i=1}^{n} \frac{\partial}{\partial x_i} (F_i(V) - F_i(U))\, dx\, dt =$$

$$\sum_{i=1}^{n} (\eta''(U) \frac{\partial U}{\partial x_i}, F_i(V) - F_i(U)) - \int_{\Sigma_T} (\eta'(U)U, \sum_{i=1}^{n} \nu_i (F_i(V) - F_i(U)))\, d\sigma .$$

Maintenant, dans le premier terme du second membre de (II.42) on utilise la relation de compatibilité de l'entropie pour obtenir

$$\text{(II.44)} \qquad (\eta''(U) F_i'(U) \frac{\partial U}{\partial x_i}, V - U) = ((F_i'(U))^* \eta''(U) \frac{\partial U}{\partial x_i}, V - U) =$$

$$= (\eta''(U) \frac{\partial U}{\partial x_i}, F_i'(U)(V - U)).$$

Ainsi de (II.42) et (II.43) résulte la relation :

$$\text{(II.45)} \qquad \int_{C_T} \frac{\partial}{\partial t} (\eta'(U), V - U)\, dx\, dt = \int_{C_T} \sum_{i=1}^{n} (\eta''(U) \frac{\partial U}{\partial x_i}, F_i(V) - F_i(U) -$$

$$- F_i'(U)(V - U))dx\, dt + \int_{\Sigma_T} (\eta'(U)U, \sum_{i=1}^{n} \nu_i (F_i(V) - F_i(U)))d\sigma .$$

Comme les fonctions F_i sont régulières, on déduit de la formule de Taylor la majoration

$$\text{(II.46)} \qquad (\sum_{i=1}^{n} \eta''(U) \frac{\partial U}{\partial x_i}, F_i(V) - F_i(U) - F_i'(U)(V - U)) \le C(U, \nabla U)\, |U - V|^2 .$$

Dans le second membre de (II.46) $C(U,\nabla U)$ désigne une constante qui dépend de la norme de U et de ∇U dans $L^{\infty}(\mathbb{R}^n) \times]0,T^*[$, tandis que $|.|$ désigne toujours la norme euclidienne dans $\mathbb{R}^m$.

Ainsi en réunissant les relations (II.41), (II.43) et (II.46) on obtient par intégration sur C_T :

$$\text{(II.47)} \quad \int_{|x|\leq R-MT} (\eta(V) - \eta(U) - \eta'(U)(V-U))(x,T)\,dx$$

$$\leq \int_{|x|=R} (\eta(V) - \eta(U) - \eta'(U)(V-U))(x,0)\,dx + C(U,\nabla U) \int_{C_T} |V-U|^2\,dx\,dt$$

$$- \int_{\Sigma_T} [\nu_t(\eta(V) - \eta(U) - \eta'(U)(V-U)) + \sum_{i=1}^{n} \nu_i [G_i(V) - G_i(U) - \eta'(U)(F_i(V) - F_i(U))]\,d\sigma .$$

On utilise alors la stricte convexité de la fonction $\eta(U)$ d'une part pour minorer le premier terme de (II.47

$$\gamma \int_{|x|\leq R-MT} |(V-U)(x,T)|^2\,dx,$$

et, d'autre part, pour montrer que le dernier terme de (II.47), compte tenu du signe - est négatif ; en effet, on a (d'après la relation de compatibilité de l'entropie $G_i'(U) = \eta'(U)\,F_i'(U)$) :

$$\text{(II.48)} \quad |G_i(V) - G_i(U) - \eta'(U)(F_i(V) - F_i(U))| = |G_i'(U)(V-U) - \eta'(U)F_i'(U)(V-U))$$

$$+ O(|V-U|^2)| \leq C\,|V-U|^2 \quad .$$

Ainsi en utilisant la stricte convexité de η il vient :

$$\text{(II.49)} \quad \nu_t(\eta(V) - \eta(U) - \eta'(U)(V-U)) - \sum_{i=1}^{n} \nu_i(G_i(V) - G_i(U) - \eta'(U)(F_i(V) - F_i(U))$$

$$\geq (\nu_t\gamma - C\sum_{i=1}^{n} |\nu_i|)\,|V-U|^2 \quad .$$

Bien entendu si la pente M du cône C est assez grande, ce dernier terme est positif et on a, en utilisant (II.47) et (II.49) :

$$\text{(II.50)} \quad \int_{|x|\leq R-MT} |U(x,T) - V(x,T)|^2 dx \leq C \int_0^T \int_{|x|\leq R-MT} |U(x,t) - V(x,t)|^2 dx\,dt$$

$$+ D \int_{|x|\leq R} |U(x,0) - V(x,0)|^2 dx$$

et le résultat se déduit du lemme de Gronwall.

COMMENTAIRES SUR LE PARAGRAPHE II

Bien que les théorèmes d'existence et d'unicité 1 et 2 parlent de problèmes non linéaires, ils sont par leur esprit linéaires, car ils considèrent de petites perturbations (temps petit, par exemple) de l'état linéaire. Ces théorèmes sont maintenant standards, nous les avons cités d'une part pour être complet, et d'autres part comme application de la notion de systèmes symétrisables.

On trouvera uneforme plus précise et plus détaillée du théorème d'existence dans l'exposé de Majda,(cf. aussi Klainerman et Majda [11]) dans ce volume , mais les références antérieures les plus classiques sont probablement les articles de Kato [17] , Marsden [31] et Da Veiga et Valli [9].

L'introduction de la notion d'entropie pour les systèmes hyperboliques est due à Friedrichs et Lax [12]. Il est en particulier démontré dans [12] que l'existence d'une entropie convexe implique que le système est symétrisable. Inversement, le fait pour un système d'être symétrisable, n'implique pas forcément l'existence d'une entropie convexe ; les systèmes considérés par Majda dans ce volume sont symétrisables mais n'admettent pas forcément d'entropies convexes.

Donnons deux exemples classiques, le premier qui sera étudié au § III est le p système :

$$\frac{\partial u}{\partial t} + \frac{\partial}{\partial x} p(v) = 0 \quad , \quad \frac{\partial v}{\partial t} - \frac{\partial u}{\partial x} = 0.$$

En multipliant la première équation par u et la seconde par $-p(v)$, on voit que ce système admet pour entropie :

$$\eta(u,v) = \frac{1}{2} |u|^2 + \int_{v_o}^{v} p(s)\, ds$$

et pour flux d'entropie

$$q(u,v) = p(v)u \ .$$

Si $p'(s)$ est négative (ce qui correspond à l'hyperbolicité) η est une entropie strictement convexe.

Le deuxième exemple est celui des fluides compressibles isothermes à n dimension d'espace. Les équations sont :

$$\frac{\partial}{\partial t}(\rho u) + \sum_{i=1}^{n} \frac{\partial}{\partial x_i}(u_i \rho u) + \nabla p(\rho) = 0$$

$$\frac{\partial \rho}{\partial t} + \nabla \cdot (\rho u) = 0 .$$

On pose $w(\rho) = \int_{\rho_o}^{\rho} \frac{p(s)}{s^2}\, ds$, et on multiplie scalairement la première équation par u et la seconde par $w(\rho) + \rho \frac{\partial w}{\partial \rho}$ pour obtenir par addition l'expression :

$$\frac{\partial}{\partial t}\left(\rho \frac{|u|^2}{2} + \rho w(\rho)\right) + \sum_{i=1}^{n} \frac{\partial}{\partial x_i}\left(\rho u_i \frac{|u|^2}{2} + u_i p(\rho) + \rho w(\rho) u_i\right) = 0$$

où on reconnaît entropie et flux d'entropie. Il faut cependant remarquer que l'entropie

$$\eta(u,\rho) = \rho \frac{|u|^2}{2} + \rho w(\rho)$$

n'est strictement convexe que sous l'hypothèse $|u|^2 \leq p'(\rho)/\rho$, alors que le système est toujours symétrisable.

Le théorème 3 n'est pas fondamentalement difficile mais il est intéressant car il fait jouer un rôle important à l'entropie pour les solutions discontinues.

III.- UN EXEMPLE DE SYSTÈME 2×2

À UNE DIMENSION D'ESPACE : LE P SYSTÈME

III.1.- INTRODUCTION

Il n'existe pratiquement pas de résultat concernant les solutions faibles des systèmes hyperboliques à plus d'une dimension d'espace ; par contre de nombreux progrès ont été obtenus pour les systèmes à une dimension d'espace. De manière très intuitive cela peut se justifier par les arguments suivants

(i) à une dimension d'espace le système s'écrit

$$\frac{\partial U}{\partial t} + F'(U)\,\frac{\partial U}{\partial x} = 0$$

il y a une seule matrice $F'(U)$ qui a toutes ses valeurs propres réelles on peut donc essayer de la diagonaliser pour se ramener à des équations de Transport. C'est en gros ce que l'on fait lorsque pour un système 2×2 on introduit les invariants de Riemann.

(ii) Pour un certain nombre de problèmes, on peut décrire les propriétés qualitatives des solutions faibles en observant les ondes qui se propagent vers la droite ou vers la gauche, une telle classification est bien entendue impossible à plus d'une dimension d'espace.

Une seconde simplification importante apparaît si l'on ne considère que des systèmes à deux inconnues. En particulier, on verra que pour de tels systèmes, on peut toujours trouver des invariants de Riemann et une infinité d'entropies, ce qui est lié au fait suivant : en dimension deux un champ de vecteurs admet toujours un facteur intégrant et ceci n'est plus vrai en dimension supérieure.

Enfin, pour obtenir des expressions explicites simples, nous allons nous limiter au p système. Encore une fois, bien qu'il s'agisse d'un cas particulier, on est en présence d'un problème physique qui nécessite l'introduction de méthodes très sophistiquées.

III.2.- GENERALITES SUR LE p SYSTEME

On considère conc le système

$$\text{(III.1)} \qquad \frac{\partial v}{\partial t} - \frac{\partial u}{\partial x} = 0, \qquad \frac{\partial u}{\partial t} + \frac{\partial}{\partial x}\,p(v) = 0\ .$$

En supposant u et v régulières et en multipliant la première équation par p(v), la seconde par u, on obtient :

(III.2) $\frac{\partial}{\partial t}\left(\frac{u^2}{2} - P(v)\right) + \frac{\partial}{\partial x}(p(v) \cdot u) = 0$

où P désigne une primitive de p. La fonction $\eta(u,v) = \frac{u^2}{2} - P(v)$ est donc une entropie du système et sa dérivée seconde est la matrice

$\begin{pmatrix} 1 & 0 \\ 0 & -p'(v) \end{pmatrix}$ elle sera donc strictement convexe sous la condition $p'(v) < 0$, $\forall v$ Sous cette hypothèse, le système (II.1) est symétrisable et les valeurs propres de la matrice

$$F'(u) = \begin{pmatrix} 0 & -1 \\ p'(v) & 0 \end{pmatrix}$$

sont les nombres $\pm\sqrt{-p'(v)}$.

On dira que le système est vraiment non linéaire (genuinely non linear) si $p''(v)$ ne s'annule jamais, (le cas linéaire correspond à $p''(v) = 0$).

Comme pour l'équation des ondes linéaires on peut essayer de décrire la solution de (III.1) à l'aide d'ondes se propageant (comme on va le voir cela consiste à diagonaliser le système et les vitesses des ondes seront $\pm\sqrt{-p'(v)}$). Pour cela on introduit deux fonctions $w_\pm$ destinées à satisfaire les propriétés suivantes :

(i) $(u,v) \to (w_+(u,v), w_-(u,v))$ est un isomorphisme

(ii) $w_\pm(u,v)$ sont solutions des équations

(III.3) $\frac{\partial}{\partial t} w_\pm + \lambda_\pm(u,v) \frac{\partial}{\partial x} w_\pm = 0$

L'équation (II.3) s'écrit aussi en désignant par U le vecteur des composantes (u,v) :

(III.4) $\nabla w_\pm \frac{\partial U}{\partial t} + \lambda_\pm \cdot \nabla w_\pm \frac{\partial U}{\partial x} = 0$;

soit en utilisant (III.1)

(III.5) $\nabla w_\pm \cdot (-F'(v) + \lambda_\pm I) \frac{\partial U}{\partial x} = (\lambda_\pm I - (F'(U))^*) \nabla w_\pm \cdot \frac{\partial U}{\partial x}$.

Il en résulte que les scalaires $\lambda_{\pm}$ doivent être les valeurs propres de la matrice $(F'(U))^*$ (ou ce qui revient au même de la matrice $F'(U)$) et que les vecteurs $\nabla w_{\pm}$ doivent être les vecteurs propres correspondants. Ceci donne :

$$\text{(III.6)} \qquad \lambda_{\pm} = \pm\sqrt{-\,p'(v)}, \qquad \nabla w_{\pm} = (\pm\sqrt{-\,p'(v)},\ \ 1)$$

La seconde équation de (II.6) s'intègre à vue pour donner:

$$\text{(III.7)} \qquad w_{\pm}(u,v) = u \pm \int^{v} \sqrt{-\,p'(s)}\ d$$

Les invariants de Riemann permettent en particulier de prouver l'apparition de chocs (cf. commentaires du § 1), aussi est-il nécessaire de décrire la nature de ces chocs.

On considère donc une solution (u,v) qui est discontinue le long de la courbe $x(t)$. Comme dans le cas de l'équation scalaire, la loi de conservation contient la condition de saut qui est toujours appelée relation de Rankine-Hugoniot :

$$\text{(III.8)} \qquad \nu_t[U] + \nu_x[F(U)] = 0\ .$$

Dans (III.8) ν_t et ν_x désignent les composantes de la normale à la courbe $x(t)$, ν_x est choisie positive. En posant $\sigma = x'(t)$ on déduit de (III.8) les relations

$$\text{(III.9)} \qquad \begin{cases} \sigma(v_+ - v_-) = -(u_+ - u_-) \\ \sigma(u_+ - u_-) = (p(v_+) - p(v_-)) \end{cases}$$

en éliminant σ dans (III.9) on obtient les courbes de choc qui donnent les positions de v_+ et u_+ en fonction de v_- et u_-

$$\text{(III.10)} \qquad u_+ = u_- \pm (v_+ - v_-)\sqrt{-\,\frac{(p(v_+) - p(v_-))}{v_+ - v_-}}\ .$$

Ceci correspond à deux vitesses de chocs qui sont respectivement :

$$\text{(III.11)} \qquad \sigma_1 = -\sqrt{-\,\frac{p(v_+) - p(v_-)}{v_+ - v_-}}\ , \qquad \sigma_2 = \sqrt{-\,\frac{p(v_+) - p(v_-)}{v_+ - v_-}}$$

les 1 chocs se propagent vers la gauche et
les 2 chocs se propagent vers la droite.

Comme dans le cas scalaire, il faut pour obtenir une solution physique, ne retenir que les courbes de chocs pour lesquelles les caractéristiques rentrent dans le choc, ce qui donne, sous l'hypothèse $p''(v) > 0$, les relations suivantes :

Pour les 1 chocs on doit avoir (cf. fig. III.1)

$$\text{(III.12)} \quad -\sqrt{-p'(v_-)} > -\sqrt{-\frac{p(v_+)-p(v_-)}{v_+-v_-}} > -\sqrt{-p'(v_+)}$$

et pour les 2 chocs on doit avoir

$$\text{(III.13)} \quad \sqrt{-p'(v_-)} > \sqrt{\frac{p(v_+)-p(v_-)}{v_+-v_-}} > \sqrt{-p'(v_+)} \ .$$

On en déduit que v décroît au travers d'un 1 choc ($v_- > v_+$) et croît au travers d'un 2 choc ($v_- < v_+$). Ces deux conditions sont équivalentes au fait suivant : sur un choc $\frac{\partial v}{\partial t}$ est une distribution négative.

On peut alors faire le lien avec la condition d'entropie en démontrant la

Proposition 1.- On suppose que $p(v)$ est une fonction décroissante et strictement convexe. On suppose que (v,u) est une solution du système (III.1) qui est régulière au voisinage d'une courbe de choc $x(t)$, on désigne par (v_-,u_-) et (v_+,u_+) les valeurs limites de cette solution de part et d'autre de la courbe de choc. Alors pour que le choc soit admissible, ($v^- > v^+$ pour un 1 choc et $v^- < v^+$ pour un 2 choc) il faut et il suffit que (v,u) satisfasse au voisinage de $x(t)$ l'inégalité d'entropie :

$$\text{(III.14)} \quad \frac{\partial}{\partial t}\left(\frac{u^2}{2} - P(v)\right) + \frac{\partial}{\partial x}(p(v)u) \leq 0.$$

Démonstration.- On va faire le calcul dans le cas d'un 2 choc, il est semblable dans le cas d'un 1 choc ; on désigne par $\eta(U)$ l'entropie $\eta(U) = \frac{u^2}{2} - P(v)$ et par $F(U)$ la fonction

$$F(U) = \begin{pmatrix} u \\ p(v) \end{pmatrix} \ ;$$

v_- et u_- étant donnés; v_+ et u_+ et σ sont déterminés par les relations (III.9) (III.10) et (III.11).

On introduit un paramètre $t \in [0,1]$ et on pose :

$$v(t) = v_- + t(v_+ - v_-), \quad \sigma(t) = \sqrt{-\frac{p(v(t)) - p(v_-)}{v(t) - v_-}}$$

$$u(t) = u_-(t) - \sigma(t)(v(t) - v_-) \quad , \quad U(t) = v(t), u(t)) \quad .$$

Pour que la relation (II.14) soit satisfaite au voisinage du choc, il faut et il suffit que sur le choc on ait

$$\text{(III.15)} \quad \sigma(1)(\eta(U(1)) - \eta(U(0))) - (p(v(1))u(1) - p(v(0))u(0)) \geq 0$$

$p(v)u$ est le flux d'entropie correspondant à l'entropie η aussi peut-on utiliser les deux expressions intégrales :

$$\eta(U(1) - \eta(U(0)) = \int_0^1 \eta'(U(t)) \,.\, U'(t)\, dt$$

et

$$p(v(1))\; u(1) - p(v(o))u(0) = \int_0^1 \eta'(U(t))\; F'(U(t))\; U'(t)\; dt \; .$$

On en déduit que la relation (III.15) est équivalente à la relation :

$$\text{(III.16)} \quad \sigma(1) \int_0^1 \eta'(U(t))\; U'(t)\; dt - \int_0^1 \eta'(U(t))\; F'(U(t))\; U'(t)\; dt \geq 0 \; .$$

Si on choisit pour primitive de $U'(t)$ la fonction $U(t) - U_-$, on peut intégrer (III.16) par parties sans faire apparaître de termes ni pour $t = 0$, ni pour $t = 1$ (utiliser la relation de Rankine-Hugoniot). La relation (III.16) est alors équivalente à la rélation :

$$\text{(III.17)} \quad - \int_0^1 \eta''(U(t))[U'(t), \sigma(1)(U(t) - U_-) - F(U(t)) - F(U_-)]\; dt \geq 0.$$

On remarque alors que par construction de $v(t)$ et $\sigma(t)$ on a :

$$\text{(III.18)} \quad F(U(t)) - F(U_-) = \sigma(t)(U(t) - U_-)$$

et ainsi il vient pour (III.17), l'expression :

$$- \int_0^1 (\sigma(1) - \sigma(t))\; \eta''(U(t))\; [U'(t), U(t) - U_-]\; dt \geq 0 \; .$$

On utilise maintenant les propriétés de convexité de la fonction p pour remarquer que dans le cas d'un 2 choc, on a :

$$\text{(III.19)} \quad \sigma(1) - \sigma(t) = \left(\sqrt{-\frac{p(v_+) - p(v_-)}{v_+ - v_-}} - \sqrt{-\frac{p(v(t)) - p(v_-)}{v(t) - v_-}} \right)$$

$$= \theta(t) \operatorname{sgn} (v_- - v_+)$$

où $\theta(t)$ désigne une fonction positive.

De plus, on a toujours, par un calcul explicite :

$$\text{(III.20)} \quad \eta''(U(t))[U'(t),U(t) - U_-]) = \frac{3}{2}(p(v_-) - p(v(t)) - p'(t))(v(t) - v_-))(v_+ - v_-)$$

$$= -\frac{3}{2}(v_+ - v_-)(v(t) - v_-)(p'(v(t)) + p'(\xi(t))) \quad .$$

Dans le dernier terme de (III.20) $\xi(t)$ désigne un point dépendant de t. Comme p' est négative, il résulte de (III.20) et de (III.19) que la relation (III.28) est réalisée si et seulement si on a v_- inférieur à v_+ ce qui démontre la proposition.

Remarque 1.- Les formules (III.12), (III.13) et la proposition 1 restent vraies mutatis mutandis, si la fonction $p''(v)$ est négative au lieu d'être positive, ou plus généralement si elle est de signe constant sur l'intervalle d'extrémités v_- et v_+.

III.3.- EXEMPLE DE SOLUTIONS ELEMENTAIRES DU PROBLEME DE RIEMANN

Le problème de Riemann non linéaire est l'analogue de la recherche de la solution élémentaire dans le cas linéaire, il consiste à résoudre le système

$$\frac{\partial v}{\partial t} - \frac{\partial u}{\partial x} = 0 \quad , \quad \frac{\partial u}{\partial t} + \frac{\partial}{\partial x} p(v) = 0$$

avec des données initiales constantes en dehors de zéro et discontinues en zéro (u_-,v_-), (u_+,v_+), bien entendu le problème est invariant par le changement $x \to \lambda x$, $t \to \lambda t$ et ainsi les solutions sont constantes le long des droites issues de l'origine.

Bien entendu, si u_+, v_+ sont sur les courbes de choc admissibles :

$$u_+ = u_- + (v_+ - v_-) \sqrt{-\frac{p(v_+) - p(v_-)}{v_+ - v_-}} \quad , \quad v_- > v_+$$

ou

$$u_+ = u_- - (v_+ - v_-) \sqrt{-\frac{p(v_+) - p(v_-)}{v_+ - v_-}} \quad , \quad v_- < v_+$$

les solutions sont des 1 chocs ou des 2 chocs (cf. figure III.1).

Il existe une seconde sorte de solutions élémentaires appelées ondes de détente. On les obtient en cherchant des solutions continues et dérivables (en dehors de t = 0) du système

$$\frac{\partial U}{\partial t} + F'(U)\,\frac{\partial U}{\partial x} = 0 \quad .$$

Comme on se place dans une phase de régularité, on va retrouver les invariants de Riemann introduits au § III.2.

En écrivant que U ne dépend que de la variable $\xi = \frac{x}{t}$, on obtient l'équation

$$\text{(III.21)} \quad 0 = -\frac{x}{t^2}\,U'_\xi + F'(U)U'_\xi\,\frac{1}{t} = \frac{1}{t}(-\xi I + F'(U))\,U'_\xi$$

ce qui conduit à deux types d'ondes

Les 1 détentes définies par les relations

$$\text{(III.22)} \quad \xi = -\sqrt{-p'(v)} \quad , \quad u'_\xi = \sqrt{-p'(v(\xi))}\,v'_\xi(\xi)$$

et les 2 détentes

$$\text{(III.23)} \quad \xi = \sqrt{-p'(v)} \quad , \quad u'_\xi = -\sqrt{-p'(v(\xi))}\,v'_\xi(\xi) \quad .$$

Comme on doit avoir $\xi(v_+) > \xi(v_-)$ on en déduit les relations

$$\text{(III.24)} \quad v_- < v_+ \quad \text{et} \quad u_+ = u_- + \int_{v_-}^{v_+} \sqrt{-p'(s)}\,ds$$

ou

$$\text{(III.25)} \quad v_- > v_+ \quad \text{et} \quad u_+ = u_- - \int_{v_-}^{v_+} \sqrt{-p'(s)}\,ds$$

Ce qui donne les types d'ondes décrites par les figures III.2.

Enfin, on introduit deux courbes C_1 et C_2 qui sont les réunions des courbes de choc et des courbes de détente qui se propagent vers l'avant, et des courbes de choc et de détente qui se propagent vers l'arrière C_1 est défini par

$$\text{(III.26)} \quad u_+ = \begin{cases} u_- + (v_+ - v_-)\sqrt{-\dfrac{p(v_+) - p(v_-)}{v_+ - v_-}} & \text{si } v_+ < v_- \\ u_- + \displaystyle\int_{v_-}^{v_+} \sqrt{-p'(s)}\,ds & \text{si } v_- < v_+ \end{cases}$$

et C_2 est défini par

$$\text{(III.27)} \quad u_+ = \begin{cases} u_- - \displaystyle\int_{v_-}^{v_+} \sqrt{-\,p'(s)}\, ds & \text{si } v_+ < v_- \\ u_+ - (v_+ - v_-)\sqrt{-\dfrac{p(v_+) - p(v_-)}{v_+ - v_-}} & \text{si } v_+ > v_- \end{cases}$$

On remarque que ces deux courbes forment sous l'hypothèse $p''(v) > 0$ un paramétrage du plan (v,u) ce qui permet de donner, pour tout problème de Riemann au moins une solution (nous n'envisageons pas dans ce paragraphe la question de l'unicité) satisfaisant la condition d'entropie, constante le long des droites $x = \lambda t$ et composée, soit d'une seule onde, soit de deux ondes se propageant en direction inverse.

Sur la figure III.2, on a représenté les solutions élémentaires formées d'un 1 choc, d'un 2 choc, d'une 1 détente, d'une 2 détente et une solution composée d'un 1 choc et d'une 2 détente.

L'étude des courbes de choc et de détente données par la figure III.1 permet aussi d'analyser la situation créee par la rencontre au point $(T,0)$ de deux chocs. On peut donner à ce problème une solution constante sur les droites $x = \lambda(t - T)$ et satisfaisant à la condition d'entropie. Ceci est décrit sur la figure III.3. Conformément à l'intuition, on trouve que la collision des deux chocs de direction opposée produit deux chocs de direction opposée, tandis que la collision de deux chocs de même direction produit un choc de même direction et une onde de détente de direction opposée.

III.4.- LA METHODE DE GLIMM

On se propose de décrire de manière très sommaire la méthode de Glimm et de montrer comment sa mise en œuvre conduit à des schémas numériques classiques. Il s'agit, à partir du problème de Riemann, de construire, par une méthode approchée, une solution entropique du système hyperbolique

$$\text{(III.28)} \quad \frac{\partial U}{\partial t} + \frac{\partial}{\partial x}(F(U)) = 0.$$

L'interaction entre la démonstration des résultats et l'intuition numérique a joué un rôle considérable dans l'élaboration de cette méthode (Glimm [13], 1966) ; et à quelques modifications près, c'était jusqu'à l'introduction de la méthode de Di Perna (1980), qui est décrite en détail au § IV, l'unique moyen de prouver l'existence de solution faible pour des systèmes hyperboliques. De plus cette démarche conduit à une analyse fine des solutions approchées ; cette analyse est proche de l'interprétation physique de l'interaction non linéaire entre les ondes et conduit à des résultats sur le comportement asymptotique des solutions.

Comme dans le paragraphe précédent, on se limitera chaque fois que cela simplifiera l'exposé au p système vraiment non linéaire. On désignera donc par $U = (v,u)$ la solution et par $F(U)$ la fonction $(v,u) \longmapsto (-u,p(v))$.

On introduit un pas de temps Δt et un pas d'espace h, tous deux destinés à tendre vers zéro et on considère le réseau formé des points $(nh,k\Delta t)$ (cf. fig. III.4). On suppose qu'à l'instant $2k\Delta t$ la solution approchée est constante sur les intervalles de la forme $]2nh, 2(n+1)h[$; on note U_{2n+1} cette valeur ; on résoud alors au point $(2nh, 2k\Delta t)$ un problème de Riemann avec U_+ et U_- donnés respectivement par $U_+ = U_{2n+1}$, $U_- = U_{2n-1}$. La solution obtenue est constante sur les droites $(x - 2nh) = \lambda(t - 2k\Delta t)$ elle a donc la forme d'un "éventail" ; de plus, d'après les théorèmes de vitesse finie de propagation, si $\Delta t/h$ est assez petit, plus précisément si on a

$$\text{(III.29)} \quad \frac{\Delta t}{h} \leq \sup_{\xi \in [v_+, v_-]} \sqrt{-p'(\xi)} \ ,$$

cet "éventail" est strictement contenu dans le rectangle $](2n-1)h, (2n+1)h[\times]2k\Delta t, (2k+1)\Delta t[$. En particulier, sur les extrémités latérales de ce rectangle, la solution est respectivement égale à U_- et U_+.

On peut ainsi superposer chacun des problèmes de Riemann, de manière à obtenir une solution exacte sur la bande $\mathbb{R} \times]2k\Delta t, (2k+1)\Delta t[$. On réitère ensuite le procédé en résolvant des problèmes de Riemann aux sommets impairs du réseau avec des données constantes sur les intervalles $](2n-1)h, (2n+1\ h[$. Une difficulté essentielle est que la solution obtenue déjà après la première étape n'est plus constante par morceaux. Il faudra donc réaliser une interpolation pour réitérer le procédé. On a donc la succession suivante d'opérations :

Résolution de problèmes de Riemann aux sommets pairs du réseau. Interpolation par des fonctions constantes sur les intervalles d'extrémités impaires. Résolution de problèmes de Riemann aux sommets impairs du réseau. Interpolation par des fonctions constantes sur les intervalles d'extrémités paires, et ainsi de suite ...

Comme dans toute méthode numérique, on peut dégager une notion de stabilité et une notion de consistance. On notera $U_{h,\Delta t}$ la solution approchée ainsi obtenue. La méthode est stable si on a

(III.30) $$\sup_{(x,t)} |U_{h,\Delta t}| + TV(U_{h,\Delta t}) \leq C,$$

où C désigne une constante indépendante de Δt et de h et TV la variation totale de la solution approchée $U_{h,\Delta t}$.

Dans le cas scalaire, on montre facilement (cf. § 1) que la norme du sup et la variation totale décroissent en fonction du temps. On remarque ensuite que si on intègre en posant par exemple

(III.31) $$U_{(2n+1),2k\Delta t} = \frac{1}{2h} \int_{2nh}^{2(n+1)h} U_{h,\Delta t}(x, 2k\Delta t^-)\, dx$$

on diminue à la fois la norme du sup et la variation totale.

Par contre, dans le cas vectoriel les estimations qui conduisent à (III.30) sont beaucoup plus difficiles. En analysant très précisément les solutions explicites construites par le procédé d'approximation, Glimm a montré que l'estimation (III.30) était vraie à condition que

(1) la valeur interpolée (par exemple à un instant d'ordre pair $2k\Delta t$) U $U_{(2n+1)h,2k\Delta t}$ s'obtienne en choisissant une valeur effectivement atteinte à l'instant $2k\Delta t$, par la solution calculée à l'instant précédent.

(2) la variation totale à l'instant zéro ne soit pas trop grande.

La condition (1) est fondamentalement différente de (III.31) et si elle assure la stabilité, elle peut conduire à des schémas qui ne soient absolument pas consistants. Un exemple très simple est obtenu en résolvant le problème de Riemann pour l'équation de Burger :

(III.32) $$\frac{\partial u}{\partial t} + \frac{\partial}{\partial x}(u^2/2) = 0, \quad \begin{cases} u(x,0) = 1 & \text{pour} \quad t < 0 \\ u(x,0) = 0 & \text{pour} \quad t > 0. \end{cases}$$

On sait que la solution exacte est donnée par

$$u(x,t) = 1 \quad \text{si} \quad x < t/2, \qquad u(x,t) = 0 \quad \text{si} \quad x > t/2.$$

Maintenant, si on applique la méthode de Glimm en choisissant d'interpoler par la valeur de la solution précédente au milieu de l'intervalle, on obtient (cf. fig. III.4), une solution approchée égale à 1 pour $x < \frac{h}{\Delta t}\, t$ et à zéro pour $x > \frac{h}{\Delta t}\, t$, qui n'a donc rien à voir avec la solution exacte de (III.32). L'erreur de cette méthode, qui est évidemment stable, est donc une erreur de consistance, et bien entendu ce type d'accidents peut se produire aussi pour les systèmes, pour les éviter Glimm a eu l'idée de choisir de manière aléatoire, sur l'intervalle $]2nh, 2(n+1)h[$, le point d'interpolation et ainsi il a prouvé la consistance de la méthode. Cette démarche a donné lieu à une interaction fructueuse entre le point de vue théorique et le point de vue numérique, Chorin [4,5] a utilisé cette méthode pour des problèmes de combustion en remplaçant le choix aléatoire du paramètre d'interpolation par une suite équidistribuée de valeurs, ensuite, Liu [27] a démontré que pourvu que la suite de points d'interpolation soit équidistribuée, la méthode était consistante, comme le montrait les calculs de Chorin [4,5] , l'étape suivante a été de déterminer un choix optimal pour la suite équidistribuée ce qui fut fait numériquement par Colella [1].

Il est enfin intéressant de remarquer que la méthode de Glimm ne diffère des schémas aux différences finies classiques que par l'interpolation. Les méthodes classiques (cf. Leroux [25,26] pour le cas scalaire) consistent à remplacer les dérivées en temps et en espace par les différences finies suivantes :

$$\frac{\partial U}{\partial t}(nh,k\Delta t) \simeq \frac{U_n^{k+1} - U_n^k}{\Delta t} \quad ; \quad \frac{\partial}{\partial x} F(U)(nh,k\Delta t) = \frac{1}{2h}\left(F(U_{n+1}^k) - F(U_{n-1}^k)\right)$$

et ensuite à ajouter un terme de viscosité artificielle pour assurer la stabilité et faire converger la solution numérique vers la solution satisfaisant la condition d'entropie, on écrit donc les schémas à trois points généraux sous la forme :

$$U_n^{k+1} = U_n^k - \frac{\Delta t}{2h}\left(F(U_{n+1}^k) - F(U_{n-1}^k)\right) + \frac{a_{n+1/2}^k}{2}\left(U_{n+1}^k\,U_n^k\right) - \frac{a_{n-1/2}^k}{2}\left(U_n^k - U_{n-1}^k\right) \tag{III.33}$$

Dans (III.33) les coefficients $a_{n+1/2}^k$ sont déterminés en fonction de la méthode et dépendent éventuellement des valeurs U_n^k, U_{n-1}^k, U_{n+1}^k. Par exemple en prenant $\alpha_{n+1/2} = 1$, on obtient

$$U_n^{k+1} = U_n^k - \frac{\Delta t}{2h}\left(F(U_{n+1}^k) - F(U_{n-1}^k)\right) + \frac{1}{2}\left(U_{n+1}^k - 2\,U_h^k + U_{n-1}^k\right) \tag{III.34}$$

et, dans le cas scalaire, en supposant F croissante, en prenant

$$a_{n+1/2}^k = \left(\frac{F(U_{n+1}^k) - F(U_n^k)}{U_{n+1}^k - U_n^k}\right)\frac{\Delta t}{h}$$

on obtient

(III.35) $U_n^{k+1} = U_n^k - \frac{\Delta t}{h} (F(U_n^k) - F(U_{n-1}^k))$

Le premier schéma est dit schéma de Lax-Friedrichs , il revient à discrétiser l'équation

(III.36) $\frac{\partial u}{\partial t} + \frac{\partial F}{\partial x} (u) = \frac{1}{2}(\frac{h}{\Delta t}) \, h \, \frac{\partial^2 u}{\partial x^2}$

tandis que le second est dit schéma de Goudounnov,il revient à discrétiser l'équation scalaire en tenant compte du fait que pour $F'(U) > 0$ la solution se propage de la gauche vers la droite .

Si à l'instant $2k\Delta t$ on résoud des problèmes de Riemann avec des données constantes sur les intervalles $]2nh, 2(n+1)h[$, on obtient une solution exacte dans la bande $\mathbb{R} \times [2k\Delta t, (2k+1)\Delta t[$, cette solution est constante et égale à U_{2n+1}^{2k} sur les droites verticales d'abscisse $(2n+1)h$. Ainsi si on interpole par la valeur moyenne de la solution sur l'intervalle $(2n-1)h, (2n+1)h$, on peut utiliser la formule de Green.

On désigne par Q le rectangle

$$Q =](2n-1)h, (2n+1)h[\times [2k\Delta t, (2k+1)\Delta t[$$

et on obtient :

(III.37) $$0 = \int\int_Q (\frac{\partial U}{\partial t} + \frac{\partial}{\partial x} F(U)) \, dx \, dt = - \frac{1}{2h} \int_{(2n-1)h}^{(2n+1)h} U(x,(2k+1)\Delta t) dx -$$

$$- \frac{1}{2}(U_{2n-1}^{2k} + U_{2n+1}^{2k}) + \frac{\Delta t}{2h} (F(U_{2n+1}^k) - F(U_{2n-1}^k))$$

soit

(III.38) $$U_{2n}^{2k+1} = \frac{1}{2}(U_{2n-1}^{2k} + U_{2n+1}^{2k}) - \frac{\Delta t}{2h} (F(U_{2n+1}^{2k}) - F(U_{2n-1}^{2k}))$$

qui n'est autre que le schéma de Lax Friedrichs (cf. (III.34)).

Pour retrouver le schéma de Goudounnov,on modifie légèrement le maillage introduit pour la méthode de Glimm (cf. fig. III.5). A chaque pas de temps, on suppose que les données sont constantes sur les intervalles $](n-1/2)h, (n+1/2)h[$ et on résoud des problèmes de Riemann aux points $(n+1/2)h$. Pourvu que $\Delta t/h$ soit assez petit, on a toujours une solution exacte dans les bandes $R = R_x \times]k\Delta t, (k+1)\Delta t[$. On calcule ensuite U_j^{k+1} en intégrant sur le rectangle

$](n-1/2)h,(n+1/2)h[\times]k\,t,(k+1)\Delta t[$, (cf. fig. III.5). Dans le cas scalaire, si $f'(u)$ est toujours positive, la solution se propage vers la droite et on retrouve bien, en intégrant l'équation $\frac{\partial u}{\partial t} + \frac{\partial}{\partial x} F(u) = 0$, la formule :

$$\text{(III.39)} \quad 0 = h(U_n^{k+1} - U_h^k) + \Delta t\,(F(U_n^k) - F(U_{n-1}^k), \quad ,$$

soit (III.35).

Bien entendu, pour ces schémas on ne sait en général pas prouver la stabilité, sauf dans le cas scalaire. Néanmoins, pour certains exemples, on peut prouver la convergeance en adaptant les méthodes de compacité par compensation du § IV (cf. Di Perna [11]).

III.5.- UN THEOREME D'UNICITE POUR LES SOLUTIONS FAIBLES DU p-SYSTEME

Au § II.1, on a vu un théorème d'unicité dans la classe des solutions fortes et au § II.2 un théorème d'unicité pour les solutions faibles, dans le cas où il existe une solution régulière. Ces deux théorèmes sont locaux, c'est-à-dire qu'ils garantissent la coïncidence des solutions au temps $t + t_o$ dans la boule $|x - x_o| < R$, pourvu que les deux solutions coïncident au temps t_o dans la boule $|x - x_o| < R + Mt_o$ ($M > 0$ assez grand). On va maintenant donner un théorème d'unicité pour le système dans la classe des solutions ne présentant que des chocs. Bien que particulier ce théorème est intéressant car il met en jeu la notion d'entropie par l'intermédiaire du système adjoint et il est une étape importante dans la démonstration d'un théorème plus général.

On désigne, comme dans les énoncés précédents, par C_T le tronc de cône :

$$C_T = \{(x,t) \mid 0 \le t \le T,\ |x| < R - MT\} \quad (M > 0 \text{ assez grand}) \ ;$$

on considère deux fonctions U_1 et U_2 solutions dans C_T du problème :

$$\text{(III.40)} \quad \frac{\partial U_i}{\partial t} + \frac{\partial}{\partial x} F(U_i) = 0 \qquad i = 1,2 \ .$$

On désigne par $\phi(x,t)$ une fonction régulière définie dans C_T à valeur dans $\mathbb{R}^m$ (si (III.40) est un système à m inconnues). En multipliant scalairement par ϕ, on obtient par soustraction et avec les notations usuelles :

$$\text{(III.41)}\quad 0 = \int\int_{C_T} \left(\frac{\partial}{\partial t}(U_1 - U_2) + \frac{\partial}{\partial x}(F(U_1) - F(U_2))\phi \; dx\, dt\right.$$

$$= - \int\int_{C_T} (U_1 - U_2)\left(\frac{\partial\phi}{\partial t} + \left(\int_0^1 (F'(sU_1 + (1-s)U_2)^* ds\right)\frac{\partial\phi}{\partial x}\right) dx\, dt$$

$$+ \int_{|x|\leq R-MT} (U_1(x,t) - U_2(x,T))\phi(x,T)\; dx$$

$$+ \int_\Sigma \nu_t(U_1 - U_2)\phi + \nu_x(F(U_1) - F(U_2))\,\phi\; d\sigma$$

$$- \int_{|x|\leq R} (U_1(x,0) - U_2(x,0))\,\phi\,(x,0)\; dx$$

Dans (III.41) Σ désigne la frontière latérale de C_T et (ν_t,ν_x) la normale extérieure à cette frontière dans $\mathbb{R}_x \times \mathbb{R}_t$. Si on suppose que U_1 et U_2 coïncident pour $t = 0$, le dernier terme de (III.41) est nul. Ainsi pour prouver l'unicité il suffit de prouver qu'il existe une ensemble de fonction $\phi(x,t)$ vérifiant les assertions suivantes :

(i) $\phi(.,T)$ parcourt un ensemble dense dans $L^1(-R,R)$;

(ii) $\phi(x,t)$ est nulle sur Σ ;

(iii) ϕ est solution du système :

$$\text{(III.42)}\quad \frac{\partial\phi}{\partial t} + \left(\int_0^1 F'(sU_1 + (1-s)U_2)^* ds\right)\frac{\partial\phi}{\partial x} = 0\ .$$

Le problème (III.42) avec donnée à l'instant T s'appelle problème adjoint et il est évident que sa résolution impliquera l'unicité ; mais cette résolution n'est pas évidente. La difficulté fondamentale est la suivante :
$\int_0^1 F'(sU_1 + (1-s)U_2)\, ds$ est en général une fonction bornée, mais non dérivable de x et t. Aussi pour résoudre (III.42), on est conduit à introduire des suites de fonctions U_1^ε, U_2^ε régulières et convergeant presque partout vers U_1, U_2. On peut alors résoudre le système

$$\text{(III.43)}\quad \frac{\partial\phi_\varepsilon}{\partial t} + \left(\int_0^1 F'(sU_1^\varepsilon + (1-s)U_2^\varepsilon)\, ds\right)\frac{\partial\phi_\varepsilon}{\partial x} = 0.$$

Mais, pour passer à la limite dans (III.43) il est indispensable d'obtenir des estimations uniformes sur $\frac{\partial\phi_\varepsilon}{\partial x}$.

Ce programme peut être mené à bien dans le cas du p système et on obtient le

Théorème 1.- On désigne par $U_i = (v_i, u_i)$ $(i = 1,2)$ deux fonctions mesurables et bornées dans $L^\infty(\mathbb{R}_x \times [0,T])$. On désigne par $\mathcal{N}$ l'enveloppe convexe de l'ensemble des valeurs prises dans $\mathbb{R}'_x \times [0,T]$ par $v_1(x,t)$ et $v_2(x,t)$. On suppose que pour tout $\xi \in \mathcal{N}$ on a :

$$p'(\xi) < 0, \; p''(\xi) > 0$$

et on pose $\quad M = \sup_{\xi \in N} \sqrt{-p'(\xi)}$

On suppose de plus que les fonctions v_1 et v_2 vérifient dans le tronc de cône C_T la condition d'entropie

$$\text{(III.44)} \quad \frac{\partial v_i}{\partial t} \leq C < +\infty \quad \text{(au sens des distributions)}$$

alors, si U_1 et U_2 coïncident pour $t = 0$ dans la boule $|x| \leq R$ elles coïncident pour $t = T$ dans la boule $|x| \leq R - MT$.

Démonstration.- On va utiliser le système adjoint et ainsi il suffit de prouver que pour toute donnée $\phi_T(.) = (\alpha_T(.), \beta_T(.)) \in \mathcal{D}(R - MT)$, le système

$$\text{(III.45)} \quad \begin{cases} \dfrac{\partial \alpha}{\partial t} - \dfrac{\partial \beta}{\partial x} = 0 \\ \dfrac{\partial \beta}{\partial t} + \left(\displaystyle\int_0^1 p'(sv_2 + (1-s)v_1)ds \right) \dfrac{\partial \alpha}{\partial x} = 0 \end{cases}$$

admet une solution appartenant à $C^1(\mathbb{R}_x \times]0,T[)$ à support dans C_T.

Pour cela, on remplace v_1 et v_2 par deux fonctions régularisées v_1^ε et v_2^ε, construites de manière à converger presque partout vers v_1 et v_2 et à préserver la relation

$$\text{(III.46)} \quad \frac{\partial v_i^\varepsilon}{\partial t} \leq C \qquad i = 1,2 .$$

Le problème (III.45) régularisé est hyperbolique (il est même symétrisable) et par des intégrations par parties standard sur le complémentaire de C_T, il est facile de voir que la solution a son support dans C_T, pour tout $\varepsilon > 0$. On se propose maintenant de montrer que les dérivées

$$\gamma_\varepsilon = \frac{\partial \alpha_\varepsilon}{\partial t} \quad \text{et} \quad \delta_\varepsilon = \frac{\partial \beta_\varepsilon}{\partial t}$$

restent uniformément bornées dans $L^\infty(0,T)\ ;\ L^2(\mathbb{R}_x))$, en utilisant l'équation (III.45) régularisée (remplacer v_i par v_i^ε) on en déduira un résultat analogue pour les fonctions $\frac{\partial\alpha}{\partial x}$ et $\frac{\partial\beta}{\partial x}$. On pose :

$$\text{(III.47)} \quad H_\varepsilon = \int_0^1 - p'(sv_2^\varepsilon + (1-s)v_1^\varepsilon)\, ds$$

et comme p' est une fonction strictement négative et p" une fonction strictement positive, on déduit les relations suivantes :

$$\text{(III.48)} \quad -\frac{\partial H_\varepsilon}{\partial t} = \int_0^1 p''(sv_2^\varepsilon + (1-s)v_1^\varepsilon)(s\,\frac{\partial v_2^\varepsilon}{\partial t} + (1-s)\,\frac{\partial v_1^\varepsilon}{\partial t})\, ds \le C\ ;$$

$$\frac{\partial}{\partial t}\left(\frac{1}{H_\varepsilon}\right) = -\frac{1}{H_\varepsilon^2}\,\frac{\partial H_\varepsilon}{\partial t} \le C$$

où C désigne une constante indépendante de ε. En dérivant le système régularisé il vient

$$\text{(II.49)} \quad \frac{\partial\gamma_\varepsilon}{\partial t} - \frac{\partial}{\partial x}\,\delta_\varepsilon = 0$$

$$\text{(II.50)} \quad \frac{\partial}{\partial t}\left(\frac{1}{H_\varepsilon}\,\delta_\varepsilon\right) - \frac{\partial}{\partial x}\,\gamma_\varepsilon = 0$$

On multiplie la première équation par γ_ε, la seconde par δ_ε et on intègre dans la bande $T - t \le s \le T$ et il vient :

$$\text{(III.51)} \quad \frac{1}{2}\int_{-\infty}^{\infty}(\gamma_\varepsilon^2 + \frac{1}{H_\varepsilon}\,\delta_\varepsilon^2)(x,T)\, dx - \frac{1}{2}\int_{-\infty}^{\infty}(\gamma_\varepsilon^2 + \frac{1}{H_\varepsilon}\,\gamma_\varepsilon^2)(x,T)\, dx$$
$$+ \int_{T-t}^{T}\int_{-\infty}^{\infty}\frac{\partial}{\partial t}\left(\frac{1}{H_\varepsilon}\right)(\delta_\varepsilon)^2/2\, dx\, ds = 0$$

On désigne enfin par $\rho_\varepsilon(.,t)$ la fonction $\frac{1}{2}(\gamma_\varepsilon^2 + \frac{1}{H_\varepsilon}\,\delta_\varepsilon^2)(x,T-t)$, et de (III.48) et (III.51) on déduit une inégalité de Gronwall

$$(III.52) \quad \int_{-\infty}^{\infty} \rho_\varepsilon(x,t)\,dx \le \int_{-\infty}^{\infty} \rho_\varepsilon(x,0)\,dx + \int_{(T-t)}^{T} \int_{-\infty}^{\infty} - \frac{1}{H_\varepsilon^2} \frac{\partial H_\varepsilon}{\partial t} \rho_\varepsilon(x,s)\,dx\,ds$$

$$\le c \int_0^t \int_{-\infty}^{\infty} \rho_\varepsilon(x,s)\,dx\,ds\ .$$

ce qui montre que les dérivées par rapport à t de α_ε et β_ε restent uniformément bornées dans $L^\infty(0,T\,;L^2(\mathbb{R}_x))$. Il en est de même des fonctions $\frac{\partial\alpha_\varepsilon}{\partial x}$ et $\frac{\partial\beta_\varepsilon}{\partial x}$. On peut donc passer à la limite dans le système :

$$\frac{\partial\alpha_\varepsilon}{\partial t} - \frac{\partial\eta_\varepsilon}{\partial x} = 0 \quad , \quad \frac{\partial\beta_\varepsilon}{\partial t} + \left(\int_0^1 p'(sv_2^\varepsilon + (1-s)v_1^\varepsilon)\,ds\right) \frac{\partial\alpha_\varepsilon}{\partial x} = 0.$$

et obtenir des solutions convenables de (III.45) ce qui implique l'égalité de $U_1(x,T)$ et $U_2(x,T)$ sur la boule $|x| \le R - MT$ et termine la démonstration du théorème 1.

Remarque 2.- La condition $\frac{\partial v_i}{\partial t} \le C$ (au sens des distributions) est vérifiée si les solutions U_1 et U_2 ne présentent comme singularités que des lignes de chocs distinctes car, en dehors des chocs $\frac{\partial v_i}{\partial t}$ est une fonction bornée et sur les chocs la condition d'entropie implique justement la décroissance de v. En réunissant ce théorème 1 du § III et les théorèmes 1 et 3 du § II, on a donc l'unicité de la solution du p système pourvu que les seules singularités soient des lignes de choc isolées. Le long des ondes de détente $\frac{\partial v}{\partial t}$ est borné, mais il n'est pas borné au sommet des ondes de détentes et de tels sommets peuvent exister, soit à l'instant t = 0 si les données initiales sont discontinues, soit à des temps ultérieurs comme résultats de la rencontre de deux chocs de même nature (cf. fig. III.3). Il est alors naturel de chercher l'unicité de la solution dans des classes de fonctions (v,u) qui sont solutions faibles et qui vérifient la relation $\frac{\partial v}{\partial t} \le C < +\infty$ sauf au voisinage d'un nombre fini de points de $\mathbb{R}^n \times [0,T]$. Les théorèmes 1 du § III et 1 et 3 du § II permettent alors de ramener l'étude de l'unicité globale à un problème d'unicité locale au voisinage d'un point (x_0,T_0) ou la solution génère (de par la discontinuité de la fonction $x \to U(x,T_0)$) une onde de détente de sommet (x_0,T_0). Une étude plus fine de ce type de problème de Riemann est faite dans Liu [29]. (cf. aussi Oleinik [34] et Smoller [36]) et conduit à des résultats assez généraux d'unicité.

COMMENTAIRES SUR LE § III

L'accent a été mis sur les systèmes 2×2 et plus particulièrement le p système. Donnons deux exemples classiques.

A température constante et à une dimension d'espace, le système de la dynamique des gaz, s'écrit :

$$\frac{\partial \rho}{\partial t} + \frac{\partial}{\partial x}(\rho u) = 0$$

$$\rho\left(\frac{\partial u}{\partial t} + u\frac{\partial u}{\partial x}\right) = -\frac{\partial}{\partial x}p(\rho)$$

$\rho(x,t)$ est la densité du gaz et $u(x,t)$ la vitesse de la molécule située à l'instant t au point x. p est la pression, elle dépend de la densité selon la loi $p(\rho) = \rho^{\gamma}$ $(\gamma > 2)$. Ce système s'écrit sous la forme $\frac{\partial U}{\partial t} + \frac{\partial}{\partial x}F(U) = 0$ selon les équations

$$\text{(III.53)} \quad \begin{cases} \dfrac{\partial \rho}{\partial t} + \dfrac{\partial}{\partial x}(\rho u) = 0 \\[2mm] \dfrac{\partial u}{\partial t} + \dfrac{\partial}{\partial x}\left(\dfrac{u^2}{2} + \dfrac{\gamma}{\gamma - 1}\rho^{\gamma-1}\right) = 0 \end{cases}$$

On vérifie que les valeurs propres de la matrice $F'(0)$ sont

$$\lambda = u \pm \sqrt{\gamma\rho^{\gamma-1}} \quad .$$

Ce système peut se réduire à un p système en introduisant les coordonnées lagrangiennes définies par $X = X(x,t)$, où X est solution de l'équation différentielle :

$$\frac{\partial X}{\partial t} = u(X(x,t),t), \quad X(x,0) = \int_{x_0}^{x} \frac{1}{\rho(s)}\, ds.$$

En utilisant les règles classiques de la dérivation et en posant $\rho = \frac{1}{v}$ on transforme ainsi le système (III.53) en le p système :

$$\frac{\partial v}{\partial t} - \frac{\partial u}{\partial x} = 0, \quad \frac{\partial u}{\partial t} + \frac{\partial}{\partial x}\, v^{-\gamma} = 0.$$

On voit que pour v positif, $p(v) = v^{-\gamma}$ vérifie les hypothèses : $p'(v) < 0$, $p''(v) > 0$.

Un second exemple est fourni par l'équation des ondes non linéaires à une dimension d'espace :

$$\text{(III.54)} \qquad \frac{\partial^2 \theta}{\partial t^2} - \frac{\partial}{\partial x}\left(\sigma\left(\frac{\partial \theta}{\partial x}\right)\right) = 0$$

$\theta(x,t)$ est le déplacement par rapport à la position initiale et la tension σ est une fonction strictement croissante de $\frac{\partial \theta}{\partial x}$, qui est concave en compression ($\frac{\partial \theta}{\partial x} < 0$) et convexe en expansion ($\frac{\partial \theta}{\partial x} > 0$). En posant $v = \frac{\partial \theta}{\partial x}$, $u = \frac{\partial \theta}{\partial t}$, et $p(v) = - \sigma(v)$, on transforme l'équation (III.54) en un p système hyperbolique, mais comme $\sigma(\lambda)$ est convexe pour $\lambda > 0$ et concave pour $\lambda < 0$, ce système n'est pas vraiment hyperbolique au point $v = 0$, ce qui introduit des difficultés supplémentaires.

Le problème de Riemann est particulièrement simple pour le p système vraiment non linéaire, et nous avons donné les différents types d'interactions élémentaires pour montrer en particulier la formation d'ondes de détentes. Pour des systèmes plus généraux, la résolution du problème de Riemann se complique vite. D'une part, si la dimension augmente on n'a plus que des résultats pour $U^+ - U^-$ assez petit (cf. Lax [23]), et d'autre part, si le système n'est plus vraiment non linéaire, un certain nombre de pathologies peuvent apparaître. (cf. Liu et Smoller [30] et Keyfitz et Krantzer [20]).

Nous avons établi le théorème 1 (Unicité) pour montrer l'interaction entre la condition d'entropie et le système adjoint, des résultats plus précis nécessitent comme on l'a dit l'unicité locale du problème de Riemann pour laquelle nous renvoyons à Liu [29], Oleinik [34] et Smoller [37].

Pour des systèmes plus généraux que le p système vraiment non linéaires, les différentes notions d'entropie que nous avons introduites cessent d'être équivalentes. Unecondition plus fine que la condition

$$\text{(E)} \qquad \frac{\partial}{\partial t}\eta(u) + \frac{\partial}{\partial x} q(u) \leq 0$$

a été introduite par Conlon et Liu [2] lorsque (E) n'assure plus l'unicité du problème de Riemann, on caractérise alors la solution physique comme celle pour laquelle

$$\frac{d}{dt}\int_{-\infty}^{\infty} \eta(u)$$

est minimum (au sens des distributions) (cf. également Dafermos [8]).

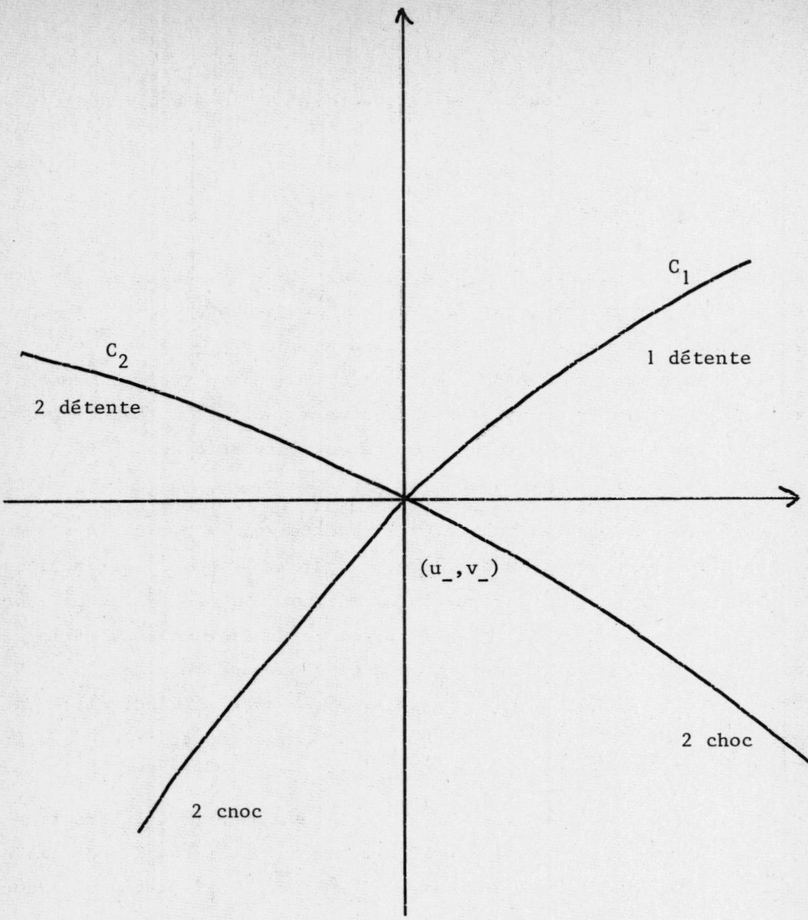
c1
1 détente
c2
2 détente
(u_,v_)
2 choc
2 cnoc

Figure III.1

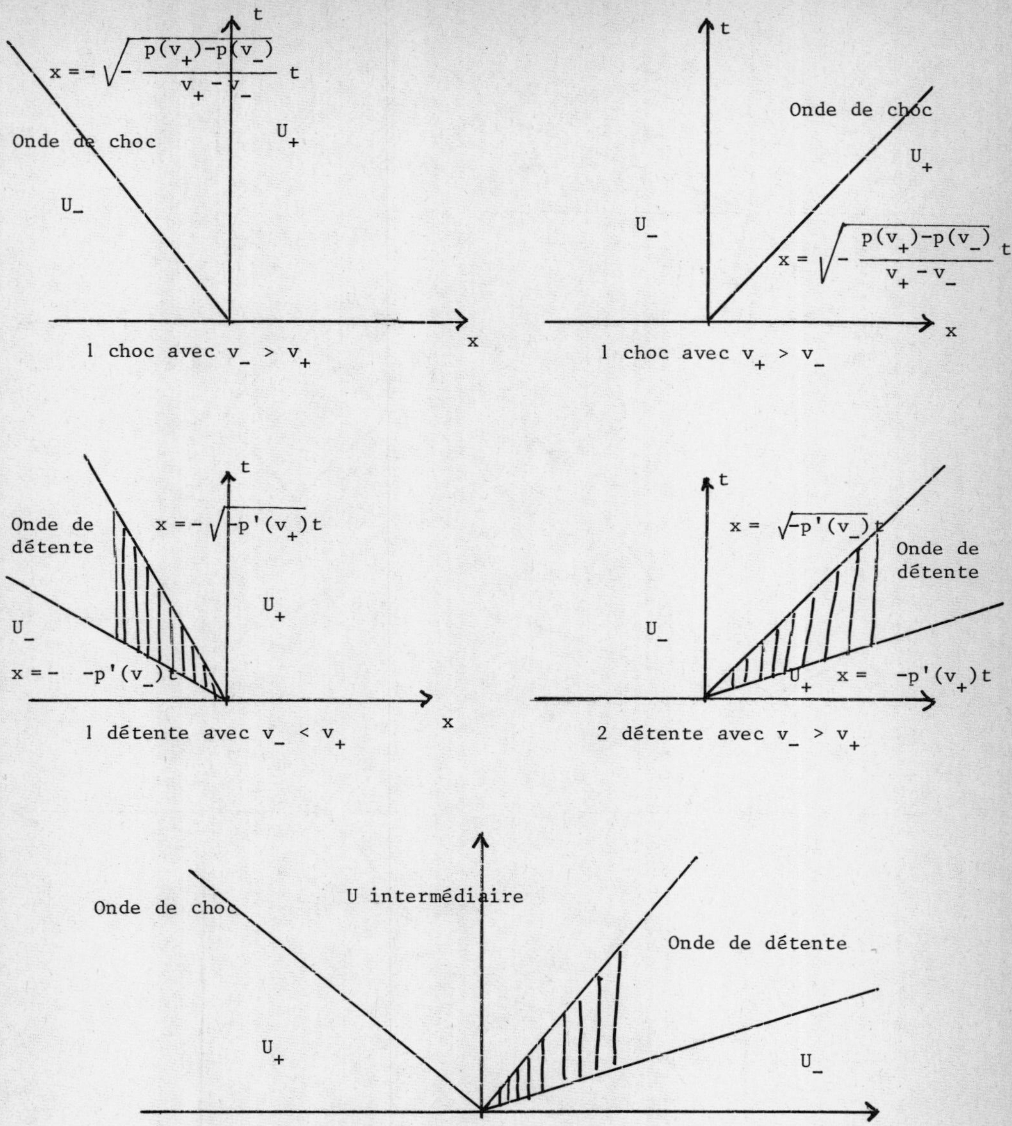

Solution générale d'un problème de Riemann

<u>Figure III.2</u>

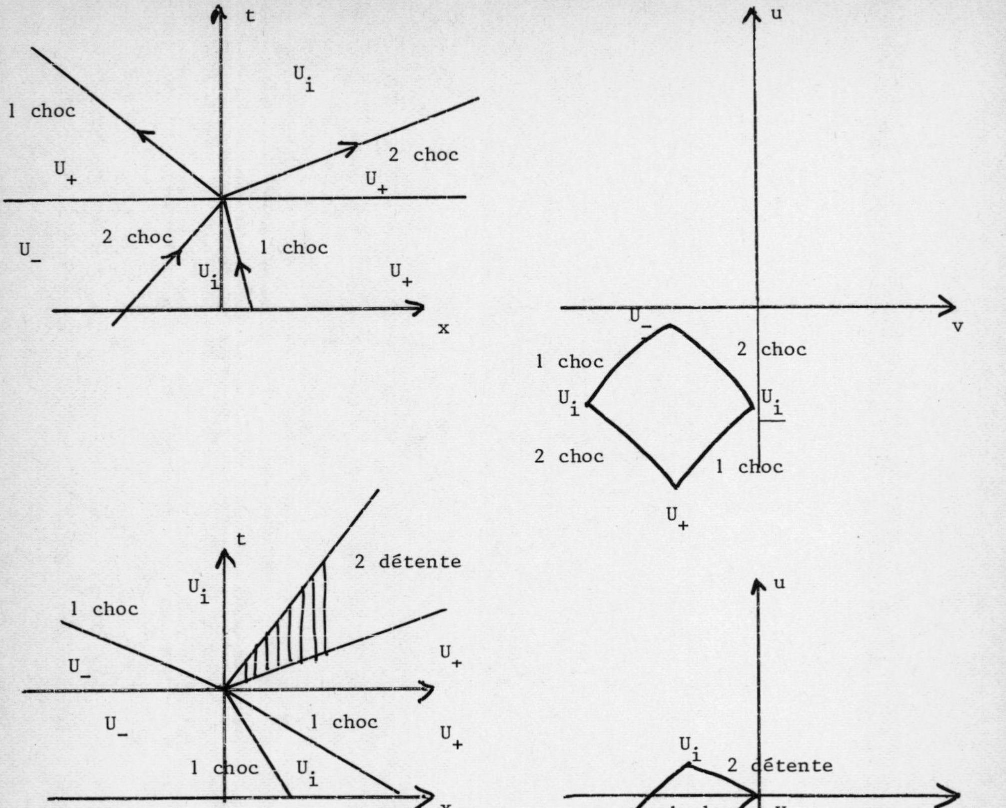

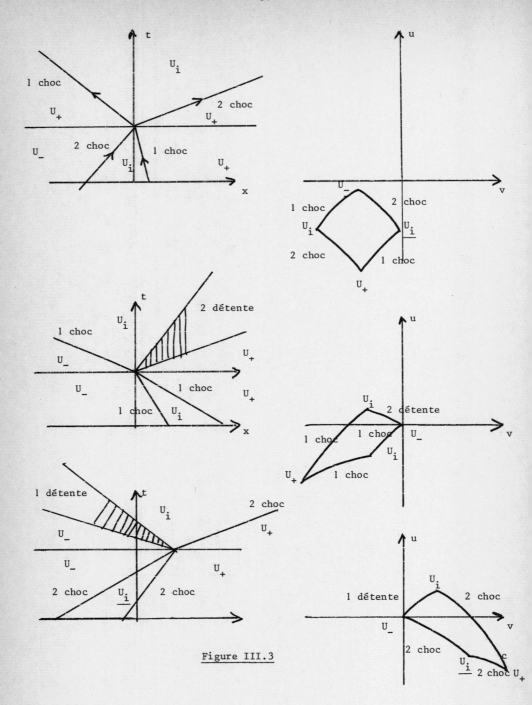

Figure III.3

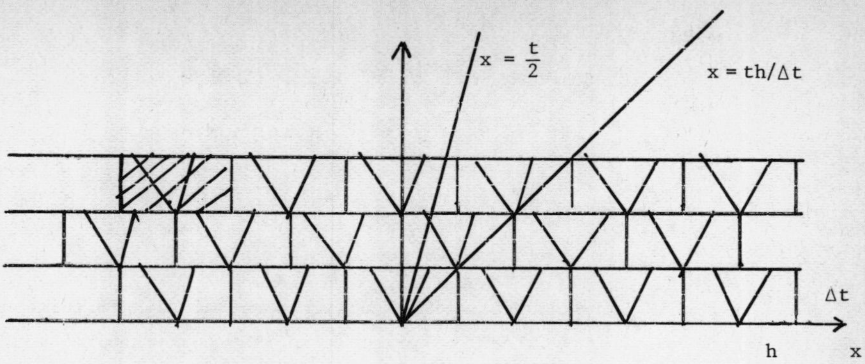

Figure III.4.- Schéma de la méthode de Glimm.$\Delta t/h$ est choisi assez petit pour que les variations de la solution restent dans les rectangles convenables. On a indiqué les deux fronts pour la solution de l'équation de Burger avec données initiales $u(x) = 1$ si $x < 0$ et $u(x) = 0$ si $x > 0$. Celui de la solution exacte et celui (faux) obtenu en interpolant aux milieux des rectangles. Le schéma de Lax-Friedrichs s'obtient en intégrant sur un rectangle élémentaire (cf. rectangle hachuré par exemple).

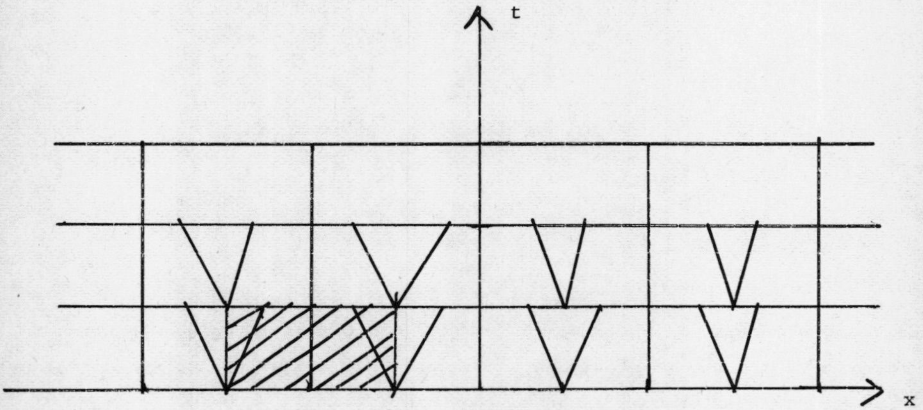

Figure III.5.- Le schéma de Goudounnov. On a représenté les problèmes de Riemann et un rectangle d'intégration.

IV.- LES ENTROPIES APPROCHÉES

ET LA CONVERGENCE VERS UNE SOLUTION

FAIBLE PAR COMPACITÉ PAR COMPENSATION

IV.1.- INTRODUCTION

On a vu que la loi de conservation scalaire admettait une infinité d'entropies tandis que pour les systèmes, il existait au plus une entropie. (évidente dansle cas du p système, mais reliée à la structure des équations physiques dans les autres cas). Pour essayer cependant d'utiliser cette notion, on va construire des entropies approchées en s'inspirant de l'approximation à hautes fréquences pour l'équation des ondes.

On utilisera ensuite ces entropies approchées pour réaliser un passage à la limite. Comme dans les paragraphes précédents, on se limitera au p système.

2.- CONSTRUCTION D'ENTROPIES APPROCHEES

On remarque que pour que $\eta(v,u)$ soit une entropie pour le système

$$\frac{\partial v}{\partial t} - \frac{\partial u}{\partial x} = 0 \quad , \quad \frac{\partial u}{\partial t} + \frac{\partial}{\partial x} p(v) = 0$$

il faut et il suffit que l'on ait (cf. § II, prop. 1) :

$$\begin{pmatrix} \frac{\partial^2\eta}{\partial v^2} , & \frac{\partial^2\eta}{\partial u\partial v} \\ \frac{\partial^2\eta}{\partial u\partial v} , & \frac{\partial^2\eta}{\partial u^2} \end{pmatrix} \begin{pmatrix} 0 & , -1 \\ p'(v) & , 0 \end{pmatrix} = \begin{pmatrix} 0 , & p'(v) \\ 1 , & 0 \end{pmatrix} \begin{pmatrix} \frac{\partial^2\eta}{\partial v^2} , & \frac{\partial^2\eta}{\partial u\partial v} \\ \frac{\partial^2\eta}{\partial u\partial v} , & \frac{\partial^2\eta}{\partial u^2} \end{pmatrix}$$

ce qui donne pour η l'équation

$$\text{(IV.1)} \qquad p'(v) \frac{\partial^2\eta}{\partial u^2} + \frac{\partial^2\eta}{\partial v^2} = 0 \ .$$

Comme $p'(v)$ est négatif, on remarque que (IV.1) n'est autre qu'une équation des ondes à coefficients variables, dont on va chercher des solutions approchées par la méthode des asymptotiques à haute fréquence.

On écrit un développement formel

$$\text{(IV.2)} \qquad \eta = e^{k\Phi}(\eta_o + \frac{1}{k}\eta_1 + \frac{1}{k^2}\eta_2 + \dots)$$

et en se reportant dans (IV.1) on obtient les équations

$$\text{(IV.3)} \qquad [p'(v)\,(\frac{\partial\Phi}{\partial u})^2 + (\frac{\partial\Phi}{\partial v})^2] = 0$$

$$\text{(IV.4)} \qquad p'(v)\,(\frac{\partial\Phi}{\partial u})(\frac{\partial\eta_n}{\partial u}) + (\frac{\partial\Phi}{\partial v})\,\frac{\partial\eta_n}{\partial v} + \frac{1}{2}\,(p'(v)\,\frac{\partial^2\Phi}{\partial u^2} + \frac{\partial^2\Phi}{\partial v^2})\,\eta_n + $$
$$+ \frac{1}{2}\,(p'(v)\,\frac{\partial^2\eta_{n-1}}{\partial u^2} + \frac{\partial^2\eta_{n-1}}{\partial v^2}) = 0\ .$$

Dans (IV.4) on convient de choisir, pour $n = 0$ le dernier terme identiquement nul. De l'équation (IV.3) on déduit que le gradient de Φ est proportionnel à $(\pm\sqrt{-p'(v)},1)$ et ainsi on retrouve pour Φ les invariants de Riemann : $\Phi_\pm(u,v) = \pm\int^v \sqrt{-p'(s)}\,ds + u$; on va poursuivre la construction en choisissant $\Phi = \Phi_-$, qui correspond à une entropie convexe (à part la convexité, le calcul est identique pour Φ_-).

On explicite l'équation (IV.4) ; on obtient à l'ordre $n \geq 1$

$$\text{(IV.5)} \qquad p'(v)\,\frac{\partial\eta_n}{\partial u} - \sqrt{-p'(v)}\,\frac{\partial\eta_n}{\partial v} + \frac{1}{4}\,\frac{p''(v)}{\sqrt{-p'(v)}}\,\eta_n + \frac{1}{2}(p'(v)\,\frac{\partial^2\eta_{n-1}}{\partial u^2} + \frac{\partial^2\eta_{n-1}}{\partial v^2}) = 0$$

et à l'ordre zéro :

$$\text{(IV.6)} \qquad p'(v)\,\frac{\partial\eta_o}{\partial u} - \sqrt{-p'(v)}\,\frac{\partial\eta_o}{\partial v} + \frac{1}{4}\,\frac{p''(v)}{\sqrt{-p'(v)}}\,\eta_o = 0\ .$$

Les équations (IV.5) et (IV.6) sont des équations de transport qui préservent la positivité éventuelle des données.

On peut donc, avec le système (IV.5) et (IV.6) construire des solutions approchées pour tout ordre fini. Enfin, on peut déduire du développement de η le développement du flux d'entropie correspondant q, en écrivant :

$$\text{(IV.7)} \qquad q(u,v) = e^{k\Phi_-}\,(q_o + \frac{q_1}{k} + \frac{q_2}{k^2} + \dots)$$

il vient, compte tenu de la relation $\Phi_- = -\int^v \sqrt{-p'(s)}\,ds + u$, les équations

$$\text{(IV.8)} \qquad \frac{\partial q}{\partial v} = p'(v)\,\frac{\partial\eta}{\partial u}\ ,\ \frac{\partial q}{\partial u} = -\,\frac{\partial\eta}{\partial v}$$

qui permettent de déterminer les coefficients b_i à partir des coefficients a_i. En particulier, on a :

$$b_o = \sqrt{-p'(v)}\ a_o , \tag{IV.9}$$

et en tenant compte de (IV.6)

$$b_1 - \sqrt{-p'(v)}\ a_1 = - \frac{1}{4} \frac{p''(v)}{p'(v)} a_o = - c(u,v) < 0. \tag{IV.10}$$

IV.3.- APPLICATION DE LA NOTION D'ENTROPIE APPROCHEE A LA CONVERGEANCE FAIBLE DES SOLUTIONS DE L'EQUATION AVEC DIFFUSION

On se propose d'utiliser simultanément la notion d'entropie approchée, et celle de compacité par compensation pour étudier la convergeance de la solution de l'équation de diffusion vers une solution entropique du p système. Nous allons donc prouver le

Théorème 1.- Soit $(v_\varepsilon, u_\varepsilon)$ une solution dans $\mathbb{R}_x \times \mathbb{R}_t^t$ de l'équation de diffusion

$$\frac{\partial v_\varepsilon}{\partial t} - \frac{\partial u_\varepsilon}{\partial x} = \varepsilon \frac{\partial^2 v_\varepsilon}{\partial x^2} \tag{IV.11}$$

$$\frac{\partial u_\varepsilon}{\partial t} + \frac{\partial}{\partial x} p(v_\varepsilon) = \varepsilon \frac{\partial^2 u_\varepsilon}{\partial x^2} . \tag{IV.12}$$

avec les conditions aux limites

$$\lim_{|x| \to \infty} (|u(x,t)| + |(v(x,t) - v\ |) = 0 \tag{IV.13}$$

et les conditions initiales

$$v_\varepsilon(x,0) = v_o(x), u_\varepsilon(x,0) = u_o(x). \tag{IV.14}$$

On suppose que $u_\varepsilon(x,t)$, $v_\varepsilon(x,t)$ restent, pour tout (x,t) et ε dans un compact $K \subset I \times J$ indépendants de ε, et que la fonction p, supposée régulière sur I, vérifie les majorations

$$p'(\xi) \leq \alpha < 0 \quad \text{et} \quad p''(\xi) \geq \beta > 0 \quad \forall\, \xi \in I. \tag{IV.15}$$

Alors on peut extraire de la famille $(v_\varepsilon, u_\varepsilon)$ une sous-famille convergeant vers

vers une solution entropique du système :

(IV.16) $$\frac{\partial v}{\partial t} - \frac{\partial u}{\partial x} = 0 \quad , \quad \frac{\partial u}{\partial t} + \frac{\partial}{\partial x}(p(v)) = 0.$$

Démonstration.- Le cœur de la démonstration consiste à extraire une sous-suite encore notée $(v_\varepsilon, u_\varepsilon)$ convergeant presque partout vers une fonction (v,u), on déduit alors de la proposition 2 du § (II.2) que (v,u) est une solution faible entropique. On note toujours U_ε le couple $(v_\varepsilon, u_\varepsilon)$.

Maintenant, en utilisant les résultats de Young [40] et Tartar [38], on sait qu'il existe une sous-famille extraite de la famille U, encore notée U_ε et une famille de mesure positive $\nu_{(x,t)}$ définies pour presque tout (x,t) possédant les propriétés suivantes : pour toute fonction G continue de $\mathbb{R}^2$ à valeur dans $\mathbb{R}$, on a

(IV.17) $$\lim_{\varepsilon \to 0} G(U_\varepsilon)(x,t) = \int_{\mathbb{R}^2} G(\lambda)\, d\nu_{x,t}(x)$$

Ainsi l'assertion U_ε converge presque partout est équivalente à l'assertion suivante : le support des mesures $\nu_{x,t}$ est toujours réduit à un point, ou ce qui revient au même, les mesures $\nu_{x,t}$ sont des masses de Dirac.

En multipliant l'équation

$$\frac{\partial v_\varepsilon}{\partial t} - \frac{\partial u_\varepsilon}{\partial v} = \varepsilon \frac{\partial^2 v_\varepsilon}{\partial x^2}$$

par $-p(v_\varepsilon)$ et l'équation

$$\frac{\partial u_\varepsilon}{\partial t} + \frac{\partial}{\partial x} p(v_\varepsilon) = \varepsilon \frac{\partial^2 u_\varepsilon}{\partial x^2}$$

par u_ε et en additionnant, on obtient la relation :

(IV.18) $$\frac{\partial}{\partial t} \int_{-\infty}^{\infty} \{ (\frac{u_\varepsilon^2}{2} - p(v_\varepsilon)) + \frac{\partial}{\partial x}(p(v_\varepsilon)u_\varepsilon) \} dx +$$

$$+ \varepsilon \int_{-\infty}^{\infty} -p'(v_\varepsilon)(\frac{\partial v_\varepsilon}{\partial x})^2 + (\frac{\partial u_\varepsilon}{\partial x})^2 \} \, dx = 0$$

d'où l'on déduit que le couple $(v_\varepsilon, u_\varepsilon)$ vérifie la majoration :

(IV.19) $$\varepsilon \int_0^\infty \int_{-\infty}^{\infty} ((\frac{\partial v_\varepsilon}{\partial x})^2 + (\frac{\partial u_\varepsilon}{\partial x})^2)\, dx \le C.$$

Soit maintenant $\eta(U) = \eta(v,u)$ une entropie quelconque et $q(U) = q(v,u)$ le flux d'entropie correspondant ; on a

$$\text{(IV.20)} \qquad \frac{\partial}{\partial t}\eta(U_\varepsilon) + \frac{\partial}{\partial x} q(U_\varepsilon) = \varepsilon \frac{\partial^2}{\partial x^2}\eta(U_\varepsilon) - \varepsilon(\eta''(U_\varepsilon)\nabla U_\varepsilon, \nabla U_\varepsilon)$$

Le second terme du second membre de (IV.20) est d'après (IV.19) une mesure bornée. Le premier terme appliqué à une fonction test ϕ s'écrit

$$\text{(IV.21)} \qquad < \varepsilon \frac{\partial^2}{\partial x^2}\eta(U_\varepsilon), \phi > = - \iint \eta'(U_\varepsilon)(\varepsilon \frac{\partial u_\varepsilon}{\partial x}) \frac{\partial \phi}{\partial x}(x,t)\, dx\, dt$$

d'après (IV.19) cette expression tend vers zéro dans l'espace $H^{-1}(\mathbb{R}^n_x \times \mathbb{R}_t)$.

Considérons maintenant deux entropies η_1, η_2 et notons q_1, q_2 les flux correspondants ; comme on a extrait de la famille U_ε une sous-famille vérifiant les relations (IV.17), on a :

$$\text{(IV.22)} \qquad \lim_{\varepsilon \to 0} \eta_i(U_\varepsilon)(x,t) = <\eta_i, \nu_{x,t}> = \bar{\eta}_i(x,t)$$

et

$$\lim_{\varepsilon \to 0} q_i(U_\varepsilon)(x,t) = <q_i, \nu_{x,t}> = \bar{q}_i(x,t) \; .$$

Mais comme toutes les expressions sont non linéaires et comme U_ε ne converge que faiblement, on n'a pas a priori de relation du type

$$\text{(IV.23)} \qquad \lim_{\varepsilon \to 0} (\eta_i(U_\varepsilon)\, q_j(U_\varepsilon))(x,t) = \bar{\eta}_i(x,t)\, \bar{q}_j(x,t) \; .$$

Aussi introduit-on les vecteurs :

$$X_\varepsilon(U_\varepsilon) = (q_1(U_\varepsilon), \eta_1(U_\varepsilon)) \qquad \text{et} \qquad Y_\varepsilon(U_\varepsilon) = (-\eta_2(U_\varepsilon), q_2(U_\varepsilon)) \quad .$$

On remarque que la divergence de X_ε, $\nabla \cdot X_\varepsilon$, et le rotationnel de Y_ε, $\nabla \wedge Y_\varepsilon$, possèdent tous deux les propriétés suivantes :

(i) Ils sont égaux à la somme d'une mesure bornée et d'une suite tendant vers zéro dans $H^{-1}(\mathbb{R}_x \times \mathbb{R}_t)$.

(ii) Ils sont somme de dérivées de fonctions appartenant à $L^\infty(\mathbb{R}_x \times \mathbb{R}^+_t)$.

Sous ces deux hypothèses on démontre à l'aide d'un théorème de Murat [32] qu'ils appartiennent à un ensemble compact de $H^{-1}_{loc}(\mathbb{R}_x \times \mathbb{R}^+_t)$. Prendre la divergence consiste en Fourier à effectuer l'opération $k \cdot \hat{X}_\varepsilon$ tandis que prendre le rotationnel consiste à faire le produit vectoriel de $\hat{Y}_\varepsilon$ avec k. Comme ces deux directions sont orthogonales on peut appliquer un résultat sur le produit des distributions X_ε et Y_ε . C'est ce que Tartar et Murat appellent la compacité par compensation. Plus précisément, on démontre (Tartar [38]) que sous l'hypothèse

(IV.24) $$\begin{cases} \nabla \cdot X_\varepsilon \in \text{ compact de } H^{-1}_{loc}(\mathbb{R}_n \times \mathbb{R}^+_t) \\ \nabla \wedge Y_\varepsilon \in \text{ compact de } H^{-1}_{loc}(\mathbb{R}_x \times \mathbb{R}^+_t) \end{cases}$$

on a, au sens des distributions

(IV.25) $$\lim X_\varepsilon \cdot Y_\varepsilon = \lim X_\varepsilon \cdot \lim Y_\varepsilon .$$

(IV.25) sécrit aussi sous la forme :

(IV.26) $$<\eta_1,\nu_{x,t}> <q_2,\nu_{x,t}> - <\eta_2,\nu_{x,t}> <q_1,\nu_{x,t}>$$

$$= <\eta_1 q_2 - \eta_1 q_2,\nu_{x,t}> .$$

Ainsi on a démontré que la famille de mesures ν "commute" **avec** les entropies. On va, à l'aide de l'approximation à haute fréquence des entropies, en déduire que les mesures $\nu_{x,t}$ sont des masses de Dirac. Comme U_ε appartient au compact K de $\mathbb{R}^2$, le support de $\nu_{x,t}$ est dans ce compact dont on choisit un nouveau système de coordonnées défini par les invariants de Riemann

(IV.27) $$w_\pm(u,v) = u \pm \int_{v_o}^{v} \sqrt{-p'(s)}\, d\sigma ,$$

et on dénote par R le plus petit rectangle qui contient le support de $\nu_{x,t}$ ((x,t) est désormais fixé) :

$$R = \{(u,v) \mid w_+^- \le w_+(u,v) \le w_+^+ \ ; \ w_-^- \le w_-(u,v) \le w_-^+\}$$

Comme l'application $(u,v) \to (w_+,w_-)$ est une bijection, il suffit de prouver les relations :

(IV.28) $$w_-^- = w_-^+ \quad \text{et} \quad w_+^- = w_+^+ .$$

On ne traitera que la premiere en désignant par $w(u,v)$ la fonction $w_-(u,v)$; la méthode est identique pour la seconde. On suppose donc que l'on a $w_- = w_-^- < w_+ = w_+^+$, et on va à l'aide de la relation (IV.26) et de la notion d'entropie approchée, s'efforcer d'obtenir une contradiction.

On désigne par η_k l'approximation à l'ordre i introduite au § 4.1 :

(IV.29) $\eta_k = e^{kw}(a_o + \frac{1}{k} a_1)$

et on note q_k le flux d'entropie correspondant.

On choisit a_o de manière à ce qu'il soit solution de (IV.6), à support compact en v et strictement positif sur R, on choisit a_1 solution de (IV.5). On désigne ensuite par $\tilde{\eta}_k$ une fonction solution de l'équation

(IV.30) $$p'(v) \frac{\partial^2 \tilde{\eta}_k}{\partial u^2} + \frac{\partial^2 \tilde{\eta}_k}{\partial v^2} = 0$$

qui coïncide à l'ordre 1 en u avec η_k sur une droite quelconque u = a :

(IV.31) $$\tilde{\eta}_k(a,v) = \eta_k(a,v) \ ; \ \frac{\partial \tilde{\eta}_k}{\partial u}(a,v) = \frac{\partial \eta_k}{\partial u}(a,v) \ .$$

Comme $\tilde{\eta}_k$ est une solution de (IV.30), c'est une entropie exacte ; on notera $\tilde{q}_k$ le flux correspondant. Comme η_k et q_k sont des solutions approchées de l'équation des ondes à l'ordre 1 et comme η_k et $\tilde{\eta}_k$ vérifient, pour le problème de Cauchy, les mêmes données initiales, on a uniformément sur K

(IV.32) $$|\eta_k - \tilde{\eta}_k| + |q - \tilde{q}_k| = e^{kw} \, O(\frac{1}{k^2}).$$

Pour distinguer les lignes de niveau de w on introduit des mesures de probabilité qui chargent les extrémités w^- et w^+ du support de ν. En extrayant une sous-famille encore notée k de la suite des $k \to \pm\infty$, on définit deux mesures $\mu^{\pm}$ par la formule :

(IV.33) $$<\mu^{\pm},h> = \lim_{k \to \pm\infty} <\nu,h\eta_{\pm k}> / <\nu,\eta_{\pm k}> \ .$$

Les supports de μ^+ et μ^- sont respectivement les extrémités

$$R \cap \{(v,u) \mid w = w^+\} \quad \text{et} \quad R \cap \{(v,u) \mid w = w^-\}$$

du rectangle (dans le plan w_-,w_+) contenant le support de ν.

On pose

(IV.34) $$\lambda^{\pm} = <\mu^{\pm}, \sqrt{-p'(v)}> \ .$$

On écrit la relation (IV.26) avec une entropie fixe de flux q et l'entropie $\tilde{\eta}_k$, $\tilde{q}_k$, il vient :

(IV.35) $$<\nu,q> - <\nu,\eta> \frac{<\nu,\tilde{q}_k>}{<\nu,\tilde{\eta}_k>} = \frac{<\nu, q\tilde{\eta}_k - \tilde{q}_k\eta>}{<\nu,\tilde{\eta}_k>}$$. Compte tenu de (IV.32)

on peut dans (IV.35) remplacer $\tilde{q}_k$ et $\tilde{\eta}_k$ par q_k,η_k afin de passer à la limite. Comme on a :

(IV.36) $$q_k = (\sqrt{-p'(v)} + O(1/k))\eta_k ,$$

en faisant tendre k vers $\pm\infty$ on déduit de (IV.35) les relations

(IV.37) $$<\mu^{\pm},q - \sqrt{-p'(v)}\eta> = <\nu,q - \lambda^{\pm}\eta>$$

On va maintenant prouver la relation $\lambda^+ = \lambda^-$ d'où résultera la relation :

(IV.38) $$<\mu^+,q - \sqrt{-p'(v)}\eta> = <\mu^-,q - \sqrt{-p'(v)}\eta>$$

On introduit dans (IV.26) les entropies $\tilde{\eta}_k$ et $\tilde{\eta}_{-k}$ (au lieu de η_k et η_{-k}) et on obtient

(IV.39) $$\frac{<\nu,\tilde{q}_k>}{<\nu,\tilde{\eta}_k>} - \frac{<\nu,\tilde{q}_{-k}>}{<\nu,\tilde{\eta}_{-k}>} = \frac{<\nu,\tilde{q}_k\tilde{\eta}_{-k} - \tilde{q}_{-k}\tilde{\eta}_k>}{<\nu,\tilde{\eta}_k><\nu,\tilde{\eta}_{-k}>} .$$

Pour passer à la limite, on peut remplacer les entropies et flux exacts $\tilde{q}_{\pm k}$, $\tilde{\eta}_{\pm k}$, par les entropies et flux approchés $q_{\pm k}$, $\eta_{\pm k}$.

En utilisant (IV.36) on voit que le premier membre de (IV.39) converge vers $\lambda_-^+ - \lambda_-^-$. D'autre part les minorations

$$<\nu,\eta_k> \geq C\, e^{k(w^+-\varepsilon)}, \quad <\nu,\eta_{-k}> \geq C\, e^{-k(w^- -\varepsilon)}$$

permettent de montrer que le second membre de (IV.39) tend vers zéro, ce qui conduit à l'égalité $\lambda_+ = \lambda_-$ et donc à la relation (IV.38). Dans la relation (IV.38) on remplace q,η par $\tilde{q}_k$ et $\tilde{\eta}_k$ et on obtient :

(IV.40) $$<\mu^+, \sqrt{-p'(v)}\,\tilde{\eta}_k - \tilde{q}_k> = <\mu^-, \sqrt{-p'(v)}\tilde{\eta}_k - \tilde{q}_k>$$

Le premier membre de (IV.40) s'écrit

$$<\mu^+, \sqrt{-p'(v)}\tilde{\eta}_k - \tilde{q}_k> = <\mu^+, \sqrt{-p'(v)}\eta_k - q_k + e^{kw}O(\tfrac{1}{k^2})>$$

$$= <\mu^+, e^{kw}/k\;\; c(u,v)> + <\mu^+; e^{kw}/k\; O(\tfrac{1}{k^2})>$$

$c(u,v)$ est la fonction qui apparaît dans (IV.10), elle est strictement positive, c'est ici qu'intervient l'hypothèse de vraie non linéarité. Ainsi pour k assez grand, le premier membre de (IV.40) est minoré par $C_1 e^{kw_+}/k$, tandis que, par un calcul analogue, le second est majoré par $C_2 e^{kw_-}/k$ ce qui contredit l'hypothèse $w_+ > w_-$ et termine la démonstration. □

Remarque 1.- Le fait que $p''(v)$ soit positif n'a pas été essentiel ; ce qui a été essentiel est le fait que $p''(v)$ ne s'annule pas. En vue d'une application d'élasticité, on va étendre le résultat du théorème 1 au cas où $p'(v)$ peut s'annuler en un point. C'est l'objet du

Théorème 2.- On désigne toujours par $u_\varepsilon, v_\varepsilon$ la suite des solutions du problème perturbé

$$\text{(IV.41)}\quad \begin{cases} \dfrac{\partial v_\varepsilon}{\partial t} - \dfrac{\partial u_\varepsilon}{\partial t} = \varepsilon \dfrac{\partial^2 u_\varepsilon}{\partial x^2} \\[2ex] \dfrac{\partial u_\varepsilon}{\partial t} + \dfrac{\partial}{\partial x} p(v_\varepsilon) = \varepsilon \dfrac{\partial^2 u_\varepsilon}{\partial x^2} \end{cases} .$$

$v_\varepsilon(x,0) = u_o(x)$, $u_\varepsilon(x,0) = u_o(x)$.

On suppose que $u_\varepsilon(x,t)$ et $v_\varepsilon(x,t)$ appartiennent, pour tout $\varepsilon > 0$ à un compact $I \times J$ de $\mathbb{R}^2$; on supposera que 0 appartient à I et au lieu de l'hypothèse (IV.15) on suppose que p vérifie les relations

$$\text{(IV.42)}\quad \begin{gathered} p'(\xi) \le \alpha < 0, \quad p''(0) = 0, \quad p''(\xi) < 0 \quad \text{si } \xi \in I \text{ et } \xi > 0, \\ p''(\xi) > 0 \quad \text{si } \xi \in I \text{ et } \xi < 0. \end{gathered}$$

alors comme dans le cas du théorème 1, on peut extraire de la famille $(v_\varepsilon, u_\varepsilon)$ une sous-famille convergeant vers une solution entropique du p-système.

Démonstration.- On montre à nouveau que le support de la mesure $\nu_{x,t}$ est réduit à un point ; on désigne maintenant par

$$w = u - \int_o^v \sqrt{-\,p'(s)}\; d\sigma\,, \quad z = u + \int_o^v \sqrt{-\,p'(s)}\; d\sigma.$$

les invariants de Riemann et par R le plus petit rectangle (dans le plan des w,z) contenant le support de $\nu_{x,t}$, dans le plan des (v,u) R est un quadrilatère curviligne dont les sommets sont notés A, B, C, D (cf. fig. IV.1). Si la droite $v = 0$ ne rencontre pas R, on sait déjà que le support de ν est réduit à un point, on doit donc seulement considérer le cas où R rencontre l'axe $v = 0$. Les formules

$$\text{(IV.43)} \qquad b_1 - \sqrt{-p'(v)}\, a_1 = \frac{1}{4}\frac{p''(v)}{p'(v)}\, a_o$$

et

$$\text{(IV.44)} \qquad < \mu^+, \sqrt{-p'(v)}\, \tilde{\eta}_k - \tilde{q}_k > \; = \; < \mu^-, \sqrt{-p'(v)}\, \tilde{\eta}_k - \tilde{q}_k >$$

restent valables.

Si les arcs des courbes A, B et C, D sont situés strictement à gauche et à droite de l'axe $v = 0$, on en déduit que, pour k grand, les deux membres de (IV.44) sont de signe opposé, aussi la seule configuration possible correspond au cas où A et C sont sur l'axe $v = 0$ et où les supports de μ_+ et μ_- sont respectivement les points $C = (\mu^+, 0)$ et $A = (\mu_-, 0)$.

On en déduit que pour tout couple entropie, flux d'entropie, on a

$$\text{(IV.45)} \qquad q(C) - \sqrt{-p'(0)}\, \eta(C) = q(A) - \sqrt{p'(0)}\, \eta(A).$$

En refaisant le même raisonnement avec l'invariant de Riemann z au lieu de w on obtient

$$\text{(IV.46)} \qquad q(C) + \sqrt{-p'(0)}\, \eta(C) = q(A) + \sqrt{-p'(0)}\, \eta(A).$$

On en déduit que C = A ce qui termine la démonstration.

4.- LES MAJORATIONS A PRIORI UNIFORMES

Le paragraphe IV.3 a été consacré à une démonstration de convergence sous l'hypothèse que les solutions restaient uniformément bornées dans L^∞. En fait, cette hypothèse n'est pas facile à vérifier et n'est démontrée complètement que dans le cas des équations de l'élasticité.

<u>Théorème 3.-</u> <u>On suppose que la fonction</u> $p(v)$ <u>est régulière et vérifie les hypothèses suivantes</u>

$$\text{(IV.47)} \qquad p'(v) < 0, \quad p''(v)\,.\,v \le 0 \qquad \forall\, v \in R$$

<u>alors pour toute donnée initiale</u> $v(.,0)$, $u(.,0)$ <u>dans</u> $L(\mathbb{R}^2)$, <u>la solution du système</u>

$$\text{(IV.48)} \quad \begin{cases} \dfrac{\partial v_\varepsilon}{\partial t} - \dfrac{\partial u_\varepsilon}{\partial x} = \varepsilon \dfrac{\partial^2 v_\varepsilon}{\partial x^2} \\[2ex] \dfrac{\partial u_\varepsilon}{\partial t} + \dfrac{\partial}{\partial x} p(v_\varepsilon) = \varepsilon \dfrac{\partial^2 u_\varepsilon}{\partial x^2} \end{cases}$$

<u>est uniformément bornée dans</u> $L^\infty(\mathbb{R}_x \times \mathbb{R}_t^+)$.

<u>Démonstration</u>.- On introduit à nouveau les invariants de Riemann

$$w = u + \int_0^v \sqrt{-p'(s)}\, d\sigma \quad , \quad z = u - \int_0^v \sqrt{-p'(s)}\, d\sigma \ ,$$

et on pose

$$C = \sup|w(.,0)| \quad , \quad D = \sup|z(.,0)| \ .$$

La région R limitée par les courbes $w(u,v) = \pm C$, $z(u,v) = \pm D$ (cf. fig. IV.2) est un convexe borné (ceci est une conséquence immédiate des relations (IV.47)).

On introduit ensuite un champ de vecteur régulier, auxiliaire $N(u,v)$ qui sur ∂R est orienté vers l'intérieur de R. On note toujours U la fonction (v,u) et $F(U)$ la fonction $(-u,p(v))$. Au lieu du système (IV.48) on considère un système perturbé :

$$\text{(IV.49)} \quad \frac{\partial}{\partial t} U + \frac{\partial}{\partial x} F(U) = \varepsilon \Delta U + \alpha N(U)$$

où α est un nombre positif destiné à tendre vers zéro. On omet d'écrire la dépendance de U par rapport au nombre ε et α, et on va montrer que $U(x,t)$ reste toujours dans R. Des méthodes classiques pour les équations paraboliques permettront ensuite un passage à la limite (par rapport à $\alpha \to 0$) conduisant à un résultat analogue pour les solutions de (IV.48). En multipliant (IV.49) par $\nabla w(u,v)$ ou par $\nabla z(u,v)$ on obtient les deux équations :

$$\begin{cases} \dfrac{\partial}{\partial t} w - \sqrt{-p'(v)}\, \dfrac{\partial w}{\partial x} = \dfrac{\partial^2 w}{\partial x^2} + \varepsilon \dfrac{p''(v)}{2\sqrt{-p'(v)}} \left(\dfrac{\partial v}{\partial x}\right)^2 + \alpha \nabla . w . N, \\[2ex] \dfrac{\partial z}{\partial t} + \sqrt{-p'(v)}\, \dfrac{\partial z}{\partial x} = \varepsilon \dfrac{\partial^2 z}{\partial x^2} - \varepsilon \dfrac{p''(v)}{2\sqrt{-p'(v)}} \left(\dfrac{\partial v}{\partial x}\right)^2 + \alpha \nabla . z . N \ . \end{cases}$$

On considère ensuite la courbe

$$\Gamma(t) : x \to v(x,t),\ u(x,t)$$

pour t = 0. Si elle sortait de R, il y aurait un temps T où elle serait tangente à ∂R, montrons que ceci est impossible.

Supposons que ceci se produise sur $\partial R \cap \{v = C\}$ en un point $(v(x,t),u(x,t))$. En ce point $w(x,t)$ atteint un maximum absolu, par rapport à x et est supérieur à toute valeur prise par w à des instants antérieurs, on en déduit les relations

$$\frac{\partial w}{\partial x} = 0 \quad , \quad \frac{\partial^2 w}{\partial x^2} \leq 0 \quad , \quad \frac{\partial w}{\partial t} \geq 0.$$

ce qui compte tenu de la première équation de (IV.50) donne

$$\text{(IV.51)} \qquad \varepsilon \frac{p''(v)}{2\sqrt{-p'(v)}} \left(\frac{\partial v}{\partial x}\right)^2 + \alpha \nabla w \, . \, N \geq 0$$

or ceci est impossible car sur $\partial R \cap \{w = C\}$ on a $p''(v) \leq 0$ d'après l'hypothèse et $\alpha \nabla . N \leq 0$.

Ceci achève la démonstration du théorème 3.

En réunissant les théorèmes 2 et 3 on obtient maintenant le résultat suivant :

Théorème 4.- On désigne par $p(v)$ une fonction régulière vérifiant les hypothèses

$$p'(v) < 0 \quad , \quad v\,p''(v) \leq 0 \quad , \quad p''(v) \neq 0 \quad \text{si} \quad v \neq 0 ,$$

alors pour toute donnée initiale (v_o,u_o), bornée, continue par morceaux et tendant assez vite vers zéro pour $|x| \to \infty$, le p-système

$$\frac{\partial v}{\partial t} - \frac{\partial u}{\partial x} = 0 \quad , \quad \frac{\partial u}{\partial t} + \frac{\partial}{\partial x} p(v) = 0$$

admet une solution entropique, limite d'une suite de solutions du problème de régularisé

$$\frac{\partial v_\varepsilon}{\partial t} - \frac{\partial u_\varepsilon}{\partial x} = \varepsilon \frac{\partial^2 v_\varepsilon}{\partial x^2} \quad , \quad \frac{\partial u_\varepsilon}{\partial t} + \frac{\partial}{\partial x} p(v_\varepsilon) = \varepsilon \frac{\partial^2 u_\varepsilon}{\partial x^2} .$$

Remarque 2.- Les équations de la mécanique des gaz conduisent à un p système avec une fonction p de la forme $p(v) = v^{-\gamma}$ $(\gamma > 1)$ et la solution sera à valeur dans l'ensemble des fonctions (v,u) vérifiant $v(x,t) > 0$. Dans ce domaine p vérifie les relations

$$p'(v) < 0 \quad , \quad p''(v) > 0.$$

aussi ne dispose-t-on plus de toutes les conclusions du théorème 3. On désigne toujours par

$$w = u + \int_{v_o}^{v} \sqrt{- p'(s)} \cdot d \quad , \quad z = u - \int_{v_o}^{v} \sqrt{- p'(s)}\ d$$

($v > 0$, $v_o > 0$) les invariants de Riemann.

La première équation de (IV.50) conduit alors à une majoration ($p''(v) > 0$) et la seconde à une minoration , on montre ainsi que $\sup_x w(x,t)$ est une fonction croissante, tandis que le inf $z(x,t)$ est une fonction décroissante ce qui montre que la solution $(v_\varepsilon, u_\varepsilon)$ reste dans une région de la forme décrite sur la fig. IV.3. On peut en déduire les estimations :

(IV.52) $\quad v > v_* \quad ; \quad |u| \leq M.$

COMMENTAIRES SUR LE § IV

Nous avons exposé dans ce paragraphe un théorème d'existence de solution pour des systèmes hyperboliques. Le résultat complet n'est obtenu actuellement que pour des systèmes très particuliers (élasticité). Mais la méthode résultant de contributions essentielles de plusieurs auteurs (Lax [24] pour l'introduction des entropies approchées, Murat [3] ou Tartar [38] pour la compacité par compensation, Tartar [38] pour l'introduction des mesures de Young et Di Perna [11] pour une synthèse générale) est actuellement en pleine expansion. Elle permet aussi d'obtenir des résultats de convergence pour des méthodes numériques en particulier dans le cas de l'élasticité, pour les méthodes de Lax Friedrichs et de Godounnov décrites au § III.4 (cf. Di Perna [11]).

Par rapport à la méthode de Glimm, elle présente les inconvénients suivants : elle n'est actuellement valable que pour les systèmes 2 × 2, en particulier elle ne conduit pas à une étude de leur comportement asymptotique. Par contre, elle présente les avantages suivants (dans les cas où elle s'applique) elle ne nécessite pas que la variation totale soit petite et peut s'appliquer au moins à un cas qui n'est pas vraiment linéaire. Il est d'ailleurs vraisemblable que pour le p système l'hypothèse $p''(v)$ s'annule en un point puisse être bientôt remplacée par l'hypothèse $p''(v)$ ne s'annule qu'en un nombre fini de points.

La démonstration du théorème 4 reprend les idées introduites par Chueh, Conley et Smoller [6] et exposées également dans [37]. On trouve dans [6] et [37] une introduction de cette notion et des théorèmes généraux de caractérisation pour des équations d'évaluation (systèmes non linéaires hyperboliques ou paraboliques). Il est intéressant de noter qu'un résultat partiel du type de ceux donnés dans [6] avait été obtenu antérieurement par Lax [24] comme première application de la notion d'entropie approchée.

Pour le p système avec $p(v) = v^{-\gamma}$ $(\gamma > 1)$, il ne manque, en vue d'obtenir des résultats d'existence généraux qu'une majoration uniforme de v. Ceci est une difficulté fondamentale du problème, v est l'inverse de la densité et v tendant ers l'infini correspond à ρ tendant vers zéro, phénomène que l'on ne peut pas exclure, cf. Liu et Smoller [30]. Il serait peut-être possible de traiter ce cas en revenant à des coordonnées Euleriennes, mais alors le problème est la construction des entropies approchées car le système linéaire est hyperbolique dégénéré.

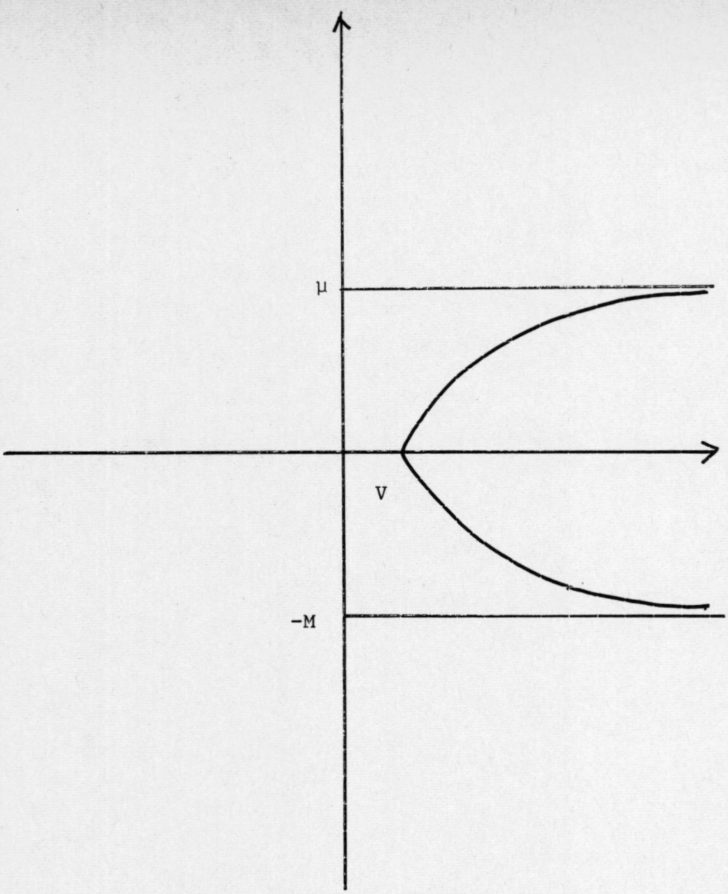

Figure IV.3

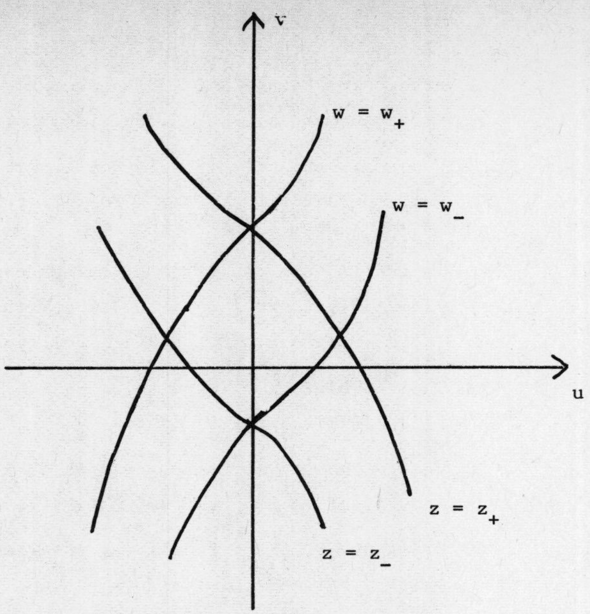

Figure IV.1

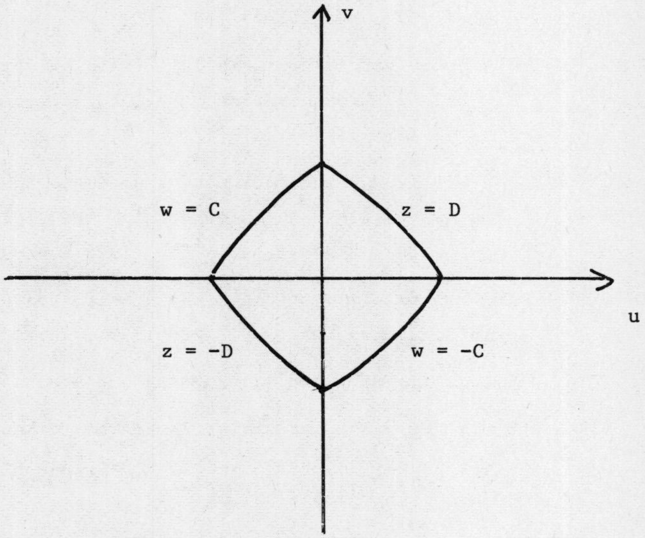

Figure IV.2

BIBLIOGRAPHIE

[1] P. COLELLA, GLIMM : Methods for Gaz Dynamic, SIAM Journal on Scientific an-Statistical Compating, Vol 3, pp. 76-110 (1982).

[2] J. CONLON and T.P. LIU : Admissibility criteria for hyperbolic conservation Law. Indiana Univ. Math. J 30 (1981) 5, pp. 641-652

[3] E.D. CONWAY et J.S. SMOLLER : Global solution of the Cauchy problem for quasilinear first order equation in several space variables. Comm. Pure Appl. Math. 19 (1966), pp. 95-105.

[4] A.J. CHORIN : Random Choice Solution of Hyperbolic System. Journ. of Comp. Phys. 22, 2 (1976), pp. 517-533.

[5] A.J. CHORIN : Random Choice Methods with applications to Reaction Gas Flow, Journ. of Comp. Phys. 25, 3 (1977), pp. 253-272.

[6] K.N. CHUEH, C.C. CONLEY et J. SMOLLER : Positively invariant Regions for systems of non linear Diffusion Equation. Indiana University Math. Jour. 26, 2 (1977), pp. 373-391.

[7] M.G. CRANDALL et T.M. LIGGETT : Generation of semi groups of non linear transformations on general Banach Spaces, Amer. J. Math. 93 (1971), pp. 265-298.

[8] C. DAFERMOS : The entropy rate admissibility criteria for solutions of hyperbolic conservation Laws. Journ. of Diff. Equations, 14, 2 (1973), pp. 202-212.

[9] M. Da VEIGA et A. VALLI : On the motion of a non homogeneous ideal incompressible fluid in an external force field, Rend. Sem. Math. Univ. Padova, vol 159, (1978), pp. 115-145.

[10] R. Di PERNA : Uniqueness of Solutions of hyperbolic conservation laws, Indiana Univ. Math. J., 28 (1979), pp. 137-187.

[11] R. Di PERNA : à paraître aux Archiv for Mech. and Analysis (1983).

[12] K.O. FRIEDRICHS et P.D. LAX : Systems of Conservation Equation with a convex Extension. Proc. Nat. Acad. Sci. U.S.A., 68, 8 (1971), pp. 1686-1688.

[13] J. GLIMM : Solution in the large for non linear hyperbolic systèmes of equations. Comm. Pure Apl. Math. 18 (1965), pp. 697-715.

[14] GOLUBITZKY et D. SCHAEFFER : Stability of schock waves for a single conservation law. Advances in Math. 16 (1975) pp. 65-71.

[15] D. HOPF : A characterisation of the blow uptime for the solution of a conservation law in several space variable. Comm. in Partial Diff. Equations, 7, 2 (1982), pp. 141-151.

[16] E. HOPF : The partial differential equation $u_t + uu_x = u_{xx}$, Comm. Pure App. Math. 3 (1950), pp. 201-230.

[17] T. KATO : The Cauchy Problem for quasilinear Symmetric Hyperbolic Systems Ard. Pub. Mech. and Anal. , 58, 3 (1975), pp. 181-205.

[18] S. KLAINERMAN et A. MAJDA : Singular limits of quasilinear Hyperbolic Systems with large parameter and the incompressible limit of compressible Fluids. Comm. Pure Appl. Math. 24 (1981), pp. 481-524.

[19] S. KALIRNEMAN et A. MAJDA : Formation of singularities for wave equations including the non linear vibrating string, Comm. Pure Appl. Math.

[20] B. KEYFITZ et H. KRANZER : A system of non strictly Hyperbolic conservation law arising in Elasticity théory. Arch. for Rab. Mech. and Anal. 72 (1979).

[21] S.N. KRUCKOV : First order quasilinear equations with several space variables, Math. USSR der Sbornik, 10 (1970), pp. 217-243.

[22] O.A. LADYZENSKAIA, SOLONNIKOV et URALTCEVA : Equations paraboliques linéaires et quasi-linéaires, Moscou (1967).

[23] P.D. LAX : Hyperbolic Systems of Conservation Laws II, Comm. Pure Appl. Math. 10 (1957), pp. 537-566.

[24] P.D. LAX : Hyperbolic Systems of Conservation Laws and the Mathematical Theory of Shock waves, S.I.A.M. Regional conference Serie in Math. 11 (1973).

[25] A.Y. LEROUX : A numerical Conception of Entropy for quasilinear equations Math. of Computation, 31, 140 (1977), pp. 848-872.

[26] A.Y. LEROUX : Numerical Stability for some equations of Gaz dynamic, Math. of Computation, 37, 156 (1981), pp. 307-320.

[27] P.L. LIONS : Generalized solutions of Hamilton Jacobi equations Pitman, Londres research Lecture Notes, n°69, Londres (1982).

[28] T.P. LIU : The deterministic version of the Glimm Scheme,Comm. in Math. Physic, 57 (1977), pp. 135-148.

[29] T.P. LIU : Uniqueness of weak solutions of the Cauchy problem for General 2 × 2 conservation Law. Jou. of Diff. Equations, 20, 2 (1976), pp. 369-388.

[30] T.P. LIU et J. SMOLLER : The vacuum state in non isentropic gas dynamics Advances in Appl. Math. 1 (1980),pp. 345-359.

[31] J. MARSDEN : Well posedness of the equation of a non homogeneous perfect fluid, Comm. in Partial Diff. Equations, 1, 3 (1976), pp. 215-230.

[32] F. MURAT : L'injection du cône positif de H^{-1} dans $W^{-1,q}$ est compacte pour tout $q > 2$. (preprint).

[33] F. MURAT : Compacité par Compensation Condition nécessaire et suffisante de continuité faible sous une hypothèse de rang constant, Ann.Scuola Norm. Sup. Pisa, 8 (1981), pp. 69-102.

[34] T. NISHIDA : Non linear hyperbolic equations and related topics in fluid dynamic, Publication Mathématiques d'Orsay, 78, 02. Université de Paris Sud, Dept. Mathématiques, Bat. 425, 91405, Orsay.

[35] O. OLEINIK : On the uniqueness of the generalized solution of Cauchy problem for a non linear system of equations occuring in Mechanics, Uspeki Math. Nauk 73 (1957), pp. 165-176.

[36] B. QUINN : Solutions with shocks an example if an L^1 contraction semi-group. Comm. Pure Appl. Math. 24 (1971), pp. 125-132.

[37] J. SMOLLER : Shock waves and reaction diffusion equations, Springer (1983).

[38] L. TARTAR : Compensated Compactness and application to partial differential equations in Research Notes on Mathematics, Non linear Analysis and Mechanics, Heriot Watt Symposium, Vol. 4, ed. R.J. KNOPS , Pitman Press (1979).

[39] A.T. VOLPERT : The spaces B.V. and quasilinear equations - Math. USSR, Sb. 2 (1967), pp. 257-267.

[40] L.C. YOUNG : Lectures on the Caculus of Variation and Optimal Control Theory, W.S. Saunders Philadelphia, Pa (1969).

SMOOTH SOLUTIONS FOR THE EQUATIONS OF COMPRESSIBLE AND INCOMPRESSIBLE FLUID FLOW

Andrew Majda*
University of California, Berkeley
Berkeley, California

1. Introduction

Here we study several topics related to the existence of smooth solutions for the general system of $m \times m$ conservation laws,

$$\frac{\partial u}{\partial t} + \sum_{j=1}^{N} \frac{\partial}{\partial x_j} F_j(u) = S(u,x,t) \tag{1.1a}$$

with the smooth initial data

$$u(x,0) = u_0(x) . \tag{1.1b}$$

Here $u = {}^t(u_1,\ldots,u_M)$ belongs to an open subset G of R^m, the state space, which arises because physical quantities such as the density or total energy should be positive and therefore the initial data $u_0(x)$ should satisfy

$$u_0(x) \in G_1 , \qquad \bar{G}_1 \subset\subset G \tag{1.2}$$

with G_1 an open subset of G. We use the notation $A_j(u) = \frac{\partial F_j}{\partial u}$, $1 \leq j \leq m$, for the corresponding $m \times m$ Jacobian matrices. The prototypical example of a system of conservation laws in R^N with $m = N+2$ is given by the compressible Euler equations,

$$\begin{aligned} &\frac{\partial \rho}{\partial t} + \operatorname{div}(\vec{m}) = 0 \\ &\frac{\partial m_i}{\partial t} + \operatorname{div}\left(\frac{m_i \vec{m}}{\rho}\right) + \frac{\partial p}{\partial x_i} = 0 , \qquad i = 1,\ldots,N \\ &\frac{\partial E}{\partial t} + \operatorname{div}\left(\vec{m}\left(\frac{E}{\rho} + \frac{p}{\rho}\right)\right) = 0 \end{aligned} \tag{1.3}$$

expressing conservation of mass, momentum, and total energy. Here ρ is the density, $\vec{m} = \rho\vec{v}$ is the momentum with ${}^t\vec{v} = (v_1,\ldots v_N)$ the fluid velocity, and $E = \frac{1}{2}\frac{\vec{m}\cdot\vec{m}}{\rho} + \rho e$ is the total energy with e the internal energy. Through thermodynamic considerations e becomes a well-defined function of the density ρ and the scalar pressure p, i.e., $e(\rho,p)$. Other interesting physical quantities, the temperature $T(\rho,p)$, and the entropy $S(\rho,p)$ are defined through the following consequence of the second law

*This work was supported in part by NSF grant MCS-81-02360 and by ARO grant 483964-25530.

of thermodynamics

$$T\,ds = de - \frac{p}{\rho^2}\,d\rho\ .$$

For simplicity in exposition, we restrict ourselves in these lectures to the special case of ideal gases where the quantities e, T, S have the explicit formulae

$$
\begin{aligned}
e(\rho,p) &= \frac{p}{\rho(\gamma-1)} = \frac{T}{\gamma-1} \\
T(\rho,p) &= \frac{p}{\rho} \qquad\qquad (1.4)\\
e^S &= p\rho^{-\gamma}\,, \qquad \gamma > 1, \text{ condtant.}
\end{aligned}
$$

In discussing solutions of (1.3) in regions of smoothness, one often uses the velocity and a convenient choice of two additional variables amongst the five quantities S, T, p, ρ, e as independent variables. For example, the Euler equations for an ideal gas can be written in terms of the variables p, v, S in regions of smoothness in the equivalent form

$$
\begin{aligned}
&\frac{Dp}{Dt} + \gamma p \operatorname{div} v = 0 \\
&\rho\,\frac{Dv}{Dt} + \nabla p = 0 \qquad\qquad (1.5)\\
&\frac{DS}{Dt} = 0
\end{aligned}
$$

with $\frac{D}{Dt} = \frac{\partial}{\partial t} + \sum_{j=1}^{3} v_j \frac{\partial}{\partial x_j}$, the convective derivative along fluid particle trajectories. Here $\rho(p,S)$ is defined by the last formula in (1.4), and since the density ρ must be positive, it is evident that the state space $G \subseteq R^5$ is defined in this example by

$$G = \{{}^t(p,v,S)\,|\,p > 0\}\ .$$

The isentropic case for (1.5) results by setting $S \equiv S_0$, a constant.

After a preliminary section on energy estimates, these lectures have three sections which we describe briefly below.

The first topic which we treat is the basic local existence theorem for the general system in (1.1). For this purpose, we introduce the integer Sobolev spaces $H^s(R^N)$, with norm

$$\|g\|_s^2 = \sum_{|\alpha| \le s} \int_{R^N} |D^\alpha g|^2\,dx \qquad (1.6)$$

and also for $u \in L^\infty([0,T],H^s)$,

$$|||u|||_{s,T} = \max_{0\le t\le T} \|u(t)\|_s\ . \qquad (1.7)$$

The basic local existence theorem for (1.1) in several space dimensions is the following result:

Theorem 2.1. Assume $u_0 \in H^s$, $s > \frac{N}{2} + 1$, and $u_0(x) \in G_1$, $\overline{G}_1 \subset\subset G$. Then there is a time interval $[0,T]$ with $T > 0$, so that the equations in (1.1) have a *unique classical solution* $u(x,t) \in C^1(R^N \times [0,T])$, with $u(x,t) \in G_2$, $\overline{G}_2 \subset\subset G$ for $(x,t) \in R^N \times [0,T]$. Furthermore,

$$u \in C([0,T],H^s) \cap C^1([0,T],H^{s-1})$$

and T depends on $\|u_0\|_s$ and G_1, i.e., $T(\|u_0\|_s, G_1)$.

Here $C([0,T],B)$ denotes the continuous functions on $[0,T]$ with values in the Banach space B. The strategy for the proof of Theorem 2.1, which we sketch at the beginning of this paper, is due independently to Kato [14] and Lax [19]. The detailed proof of Theorem 2.1 given by Kato (see [14]) uses the abstract theory of evolution equations to treat appropriate linearized problems. In fact, Kato [13] has formulated and applied this basic idea in an abstract framework which yields smooth local existence theorems for many of the interesting equations of mathematical physics.

Under somewhat different hypotheses, Kato's work has been extended recently in an interesting fashion by Crandall and Souganidis [4] to some classes of nonlinear evolution equations with non-smooth initial data and with the initial value problem posed in non-reflexive Banach spaces. At the beginning of this section, we sketch a concrete and elementary proof of Theorem 2.1 in the spirit of Lax's suggestion from [19] which only requires the elementary linear existence theory for symmetric hyperbolic systems with C^∞ coefficients [3]. Another advantage of this approach is that

(1.8) Complicated nonlinear boundary conditions can be handled in a similar fashion, including the complex free surface problems discussed in [31].

On the other hand, the abstract approach of Kato seems to have major difficulties for general nonlinear boundary value problems for hyperbolic equations because the time varying domains of generators can change in a drastic nonlinear fashion. Given Theorem 2.1, in a standard fashion, one can define a *maximal interval* of classical H^s existence $[0,T_*)$ of the solution $u \in C([0,T_*),H^s) \cap C^1([0,T_*),H^{s-1})$. After discussing Theorem 2.1, we prove the following sharp *continuation principle:*

Theorem 2.2. For any $s > \frac{N}{2} + 1$, $[0,T_*)$ with $T_* < \infty$ is a *maximal interval* of classical H^s existence for (1.1) if and only if either

$$\overline{\lim_{t \uparrow T_*}} |u_t|_{L^\infty} + |Du|_{L^\infty} = + \infty \tag{1.9}$$

or

(1.10) For *any* compact subset $K \subset\subset G$, $u(x,y)$ *escapes* K as $t \uparrow T_*$ (i.e., there exists $x_j, t_j \uparrow T_* | u(x_j,t_j) \notin K$)

where

$$|D^s u|_{L^\infty} = \sum_{|\alpha|=s} |D^\alpha u|_{L^\infty}.$$

To our knowledge, such a sharp continuation principle has not appeared explicitly in the published literature. We give a more complete discussion of Theorem 2.2 later, but we would like to point out here that the *catastrophy in* (1.9) *is associated with the formation of shock waves in the smooth solution*, while the catastrophy in (1.10) is associated with the type of blow-up familiar from O.D.E. theory. We end our general treatment of the equations in (1.1) by discussing the *uniformly local Sobolev spaces* $H^s_{u\ell}$, introduced by Kato [14] and briefly discussing variants of Theorem 2.1 and Theorem 2.2 which are valid when $u_0 \in H^s_{u\ell}$.

The equations of incompressible ideal fluid flow are the 4×4 system with unknowns (v^∞, p^∞), the velocity and scalar pressure, satisfying the (non-hyperbolic!!) system,

The Equations of Incompressible Flow

$$\rho_0 \frac{Dv^\infty}{Dt} + \nabla p^\infty = 0$$

$$\text{div } v^\infty = 0 \tag{1.11}$$

$$v^\infty(x,0) = v_0(x) ,$$

with ρ_0 a constant reference density. Both the system in (1.5) and the one in (1.11) describe fluid flow in certain regimes of motion. A basic question is the following:

How are the equations of compressible and incompressible fluids related?

This is the topic of the next section. We begin with a formal asymptotic derivation of the incompressible equations of fluid motion from the compressible equations in the limit as the *Mach number tends to zero*--this is a *singular limit* since some coefficients of the hyperbolic system in (1.1) become infinite as the Mach number tends to zero. Then we discuss a program for rigorously studying this singular limit developed recently by S. Klainerman and the author (see [15] and especially [16]), which utilizes only the elementary techniques developed in the first section. This program has three parts:

(1.12) (a) Uniform stability for the compressible solution of (1.3) independent of Mach number.

(b) Constructive existence of solutions for the incompressible equations in (1.11) by passing to the limit as the Mach number vanishes.

(c) Rigorous asymptotic expansion of the compressible Euler equations described symbolically as

$$\text{Compressible Euler} = \text{Incompressible Euler} + (\text{Linear Acoustics}) + O(M^2)$$

where M is the Mach number.

These results justify the use of linear acoustics and also rigorously indicate the discrepancies in the formal asymptotic expansion described earlier as regards $O(M)$ effects (see (3.12) and (3.15) below).

In this section we prove parts (a) and (b) of the above program in detail. In particular, following a suggestion of M. Crandall, we apply the continuation principle from Theorem 2.2 to give another simple proof of (a) and (b) which is a variant of the one presented in [16]. Next, having constructed the solution of the Euler equations in (1.11), we apply the ideas used in Theorem 2.2 to prove the following fact:

> $[0,T_*)$ with $T_* < \infty$ is a maximal interval of classical H^s existence for the incompressible Euler equations if and only if
>
> $$\overline{\lim_{t \uparrow T_*}} \, |Dv|_{L^\infty} = \infty .$$

This result improves earlier work on this equation due to Temam [25]. We remark here that the program developed in [15] and the framework of the proofs developed here for the hyperbolic case extend to the compressible Navier-Stokes equations and related viscous perturbations including the numerical method of artificial compressibility introduced independently by Chorin [28] and Temam [26]. In very interesting recent work, Klainerman and Kohn [17] have studied the incompressible limit in nonlinear elasticity using several new estimates together with the structure for the proofs from [15], [16]. We end this section by describing some interesting open problems dealing with singular limits in fluid dynamics and related topics.

Finally, in the last section we discuss a new formal asymptotic expansion due to the author [21] for the equations of combustion theory from (4.1) as the Mach number tends to zero. The result is a new system of equations describing "Zero Mach Number Combustion" - both a rigorous treatment of the limit as well as some of the mathematical properties of the equations for incompressible combustion are analyzed by P. Embid and the author in [8]. A "qualitative-quantitative" model in this spirit was used earlier with spectacular success by Ghoniem, Chorin, and Oppenheim [29] in turbulent combustion calculations in unbounded regions - here we also give an appropriate formulation for bounded regions. We end this section by specializing the formal derivation to one space dimension. If one introduces a *Lagrangian* reference frame, the equations that result are an integro-differential system of reaction diffusion equations. Under more specialized and often unrealistic assumptions ([21]) such as

(1.13)
(1) temperature deviations are small
(2) density variations are small

besides the requirement of small Mach number, many authors (see [22], for example) have found *reaction-diffusion equations in Eulerian coordinates* with the fluid dynamics decoupled and incompressible. Here, we derive a similar but more complex system in 1-D in Lagrangian coordinates under only the realistic assumptions ([21]) that

(1.14)
(1) the Mach number is small
(2) spatial pressure variations are small.

However, even in 1-D, these systems have some new features and should be studied further. In fact, since reaction-diffusion equations are a research topic with wide current interest, this is one of our main reasons for presenting a formal derivation of the appropriate system of equations for low Mach number combustion in 1-D.

The Common Structure of the Physical Systems of Conservation Laws and Friedrichs' Theory of Symmetric Systems

We first consider the case where the source terms $S(u,x,t)$ vanish. In this case, any constant u_0 belonging to the state space G is a trivial solution of (1.1). By linearizing about this smooth solution u_0 by considering solutions $u(x,t) = u_0 + \varepsilon v$, we obtain the linearized equations

$$\text{(1.15)} \qquad \begin{aligned} &\frac{\partial v}{\partial t} + \sum_{j=1}^{N} A_j(u_0)\,\frac{\partial v}{\partial x_j} = 0\,, \qquad t > 0, \quad x \in R^N \\ &v(x,0) = v_0(x)\,, \end{aligned}$$

where $A_j(u) = \frac{\partial F_j}{\partial u}$, $j = 1,\dots,N$, are the corresponding $m \times m$ Jacobian matrices. A minimum requirement for a general system of conservation laws from (1.1) is that the linearized Cauchy problem from (1.15) defines a well-posed problem. Many years ago, Friedrichs made the important observation that under reasonable conditions, almost all equations of classical physics of the form (1.1) admit the following structure: For all $u \in G$, there is a positive definite symmetric matrix $A_0(u)$ smoothly varying with u so that

(1.16) (a) $CI \le \tilde{A}_0(u) \le C^{-1} I$, $\quad \tilde{A}_0 = \tilde{A}_0^*$,
with a constant C uniform for $u \in G_1$ and any G_1 with $\overline{G}_1 \subset\subset G$

(1.16) (b) $\tilde{A}_0(u)A_j(u) = \tilde{A}_j(u)$ with $\tilde{A}_j(u) = \tilde{A}_j^*(u)$, $j = 1,\dots,N$.

For example, the equations for an ideal gas in (1.5) are symmetrized by the 5×5 matrix,

$$\tilde{A}_0(p,S) = \begin{pmatrix} (\gamma p)^{-1} & & \mathbf{O} \\ & \rho(p,S)I_3 & \\ \mathbf{O} & & 1 \end{pmatrix}. \tag{1.17}$$

Of course, symmetrizers are not unique - the reader can check that if one writes the ideal fluid equations in nonconservative form using as variables (ρ,v,T) resulting in the system,

$$\frac{D\rho}{Dt} + \rho \operatorname{div} v = 0$$

$$\rho \frac{Dv}{Dt} + \rho\nabla T + T\nabla\rho = 0$$

$$\frac{DT}{Dt} + (\gamma-1)(T)\operatorname{div} u = 0 .$$

A symmetrizer is defined by choosing $\tilde{A}_0$ as

$$\tilde{A}_0(\rho,v,T) = \begin{pmatrix} \frac{T}{\rho} & & \mathbf{O} \\ & \rho I_3 & \\ \mathbf{O} & & \frac{\rho}{(\gamma-1)T} \end{pmatrix}.$$

Next, we claim that the structural conditions in (1.16) automatically guarantee that the linearized problem is well posed through an *energy principle*. We consider a solution v of the equations

$$\begin{aligned} \tilde{A}_0(u)\frac{\partial v}{\partial t} + \sum_{j=1}^{N} \tilde{A}_j(u)\frac{\partial v}{\partial x_j} - \tilde{B}(u,x,t)v &= F \\ v(x,0) &= v_0(x) \end{aligned} \tag{1.18}$$

where we assume that $u(x,t)$ is a C^1 function with $u(x,t) \in \overline{G}_1 \subset\subset G$ and $\tilde{B}$ is a smoothly varying $m\times m$ matrix function of its arguments. We introduce the energy

$$E(t) = (\tilde{A}_0(u)v,v)$$

where

$$(v,w) = \int_{R^N} v\cdot w\,dx, \qquad \|v\|_0 = \left(\int_{R^N} |v|^2\right)^{\frac{1}{2}}$$

and also $\vec{A} = (\tilde{A}_0,\tilde{A}_1,\dots,\tilde{A}_N)$ with $\operatorname{div}\vec{A}$, the matrix defined by

$$(1.19) \qquad \operatorname{div} \vec{A} = (\tilde{A}_0)_t + \sum_{j=1}^{N} (\tilde{A}_j)_{x_j} .$$

Then using (1.18) we compute the basic *energy identity* of Friedrichs,

$$(1.20) \qquad \frac{\partial}{\partial t} E(t) = \left((\operatorname{div} \vec{A} + \tilde{B} + \tilde{B}^*)u,u\right) + 2(F,u) .$$

By applying Gronwall's inequality and the bound in (1.16a), we deduce the important stability estimate

$$(1.21) \qquad \max_{0\le t\le T} \|v\|_0 \le C^{-1} \exp\left(\tfrac{1}{2} C^{-1} |\operatorname{div} A + \tilde{B} + \tilde{B}^*|_{L^\infty} T\right)\left(\|v(0)\|_0 + \int_0^T \|f(s)\|_0 \, ds\right)$$

where $|w|_{L^\infty} = \max_{(x,t)\in R^N \times [0,T]} |w|$. We will make extensive use of estimates of the type in (1.21) in these lectures.

2. The Local Existence of Smooth Solutions for Systems of Conservation Laws

We will not discuss the uniqueness of classical solutions in $C^1(R^N \times [0,T])$ since this is proved by a standard application of the energy method described above (see [3]). In the framework we describe below, it is convenient to separate the proof of Theorem 2.1 into two parts, Theorem 2.1(a) and Theorem 2.1(b). First, we sketch the proof of

Theorem 2.1(a) Under the hypotheses of Theorem 2.1, there is a unique classical solution $u \in C^1([0,T] \times R^N)$ to the equations in (1.1) with

$$(2.1) \qquad u \in L^\infty([0,T],H^s) \cap C_w([0,T],H^s) \cap \operatorname{Lip}([0,T],H^{s-1}) .$$

Here C_w means continuity on the interval $[0,T]$ with values in the weak topology of H^s, i.e., $u \in C_w([0,T],H^s)$ means that

(2.2) For any fixed $\phi \in H^s$, $(\phi,u(t))_s$ is a continuous scalar function on $[0,T]$

where

$$(2.3) \qquad (u,v)_s = \sum_{|\alpha|\le s} \int_{R^N} D^\alpha u \cdot D^\alpha v \, dx .$$

Also $\operatorname{Lip}([0,T],H^{s-1})$ denotes the Lipschitz continuous functions on $[0,T]$ with values in the norm topology of H^{s-1}. Our proof of Theorem 2.1 is completed through the following additional fact:

Theorem 2.1(b) Any classical solution of (1.1) with $u(x,t) \in \bar{G}_2 \subset\subset G$ satisfying the regularity stated in the conclusion of Theorem 2.1(a) on some interval $[0,T]$ satisfies the additional regularity

$$u \text{ belongs to } C([0,T],H^s) \cap C^1([0,T],H^{s-1}) .$$

Remark 1. Below we give one proof of Theorem 2.1(b); S. Klainerman has recently given another different proof independently (private communication). Kato's proof via abstract evolution equations automatically yields the proof of Theorem 2.1 without first proving Theorem 2.1(a) and then Theorem 2.1(b) - this is an advantage of his approach.

The Proof of Theorem 2.1(a)

The proof proceeds via a classical iteration scheme. First, we smooth the initial data to avoid technical difficulties regarding the smoothness of the coefficients in the associated linearized problems for this iteration scheme. Thus, we choose $j(x) \in C_0^\infty(R^N)$, $\operatorname{supp} j \subseteq \{x \mid |x| \leq 1\}$, $j \geq 0$, $\int j = 1$ and set $j_\varepsilon = \varepsilon^{-N} j(\frac{x}{\varepsilon})$. We define $J_\varepsilon u \in C^\infty(R^N) \cap H^s(R^N)$ by

$$(2.4) \qquad (a) \qquad J_\varepsilon u(x) = \int_{R^N} j_\varepsilon(x-y)u(y)dy .$$

Below we use the following well-known properties of such a mollification process:

$$(2.4) \qquad (b) \qquad \text{For } u \in H^s, \quad \|J_\varepsilon u - u\|_s \to 0 \quad \text{as} \quad \varepsilon \to 0 .$$

$$(2.4) \qquad (c) \qquad \text{For } u \in H^1 \text{ and } \varepsilon \leq \varepsilon_0, \quad \|J_\varepsilon u - u\|_0 \leq \hat{C}\varepsilon\|u\|_1 .$$

Here and below $\hat{C}$ denotes a generic a priori constant which varies from relation to relation - we use $\hat{C}$ whenever the constants have no special significance in the proof. We smooth the initial data in (1.1)(b) according to the following strategy:

$$(2.5) \qquad \begin{aligned} &\text{Set } \varepsilon_k = 2^{-k}\varepsilon_0, \quad k = 0,1,2,3,\ldots, \text{ and} \\ &u_0^k(x) = J_{\varepsilon_k} u_0, \quad k = 0,1,2,3,\ldots, \end{aligned}$$

where $\varepsilon_0 > 0$ will be chosen later - these define initial data for the iteration scheme.

To construct a smooth solution of the system in (1.1), it is sufficient to differentiate the nonlinear terms and apply the symmetrizing matrix $\tilde{A}_0(u)$ from

(1.16)(a). Thus (dropping the tildas for notational convenience), we need to prove Theorem 2.1(a) for the quasi-linear symmetric system,

$$(2.6)\qquad \begin{aligned} &A_0(u)\,\frac{\partial u}{\partial t} + \sum_{j=1}^{N} A_j(u)\,\frac{\partial u}{\partial x_j} = 0 \\ &u(x,0) = u_0(x) \ . \end{aligned}$$

(We have also set $S \equiv 0$ in (2.6) for simplicity in exposition - all the proofs given below remain valid in the general case.)

We will construct the solution of (2.6) through the following iteration scheme: As a first guess, we set

$$(2.7)\quad (a)\qquad u^0(x,t) = u_0^0(x)$$

and for $k = 0,1,2,3,\ldots,$ we define $u^{k+1}(x,t)$ inductively as the solution of the linear equation

$$(2.7)\quad (b)\qquad \begin{aligned} &A_0(u^k)\,\frac{\partial u^{k+1}}{\partial t} + \sum_{j=1}^{N} A_j(u^k)\,\frac{\partial u^{k+1}}{\partial x_j} = 0 \\ &u^{k+1}(x,0) = u_0^{k+1}(x) \ . \end{aligned}$$

At the outset, it is not obvious that the iterates in (2.7)(b) are well defined - the following considerations establish this. Let G_2 be an open subset of the state space with $\overline{G}_1 \subset G_2$, $\overline{G}_2 \subset\subset G$. From the Sobolev embedding estimate for $s > \frac{N}{2}$,

$$(2.8)\qquad |v|_{L^\infty} \le C_s \|v\|_s \ ,$$

and property (2.4)(b),(c) of mollification, since $u_0(x) \in G_1$, it follows immediately that there is an $R > 0$ and a fixed choice of ε_0 so that

$$(2.9)\quad (a)\qquad \|u - u_0^0\|_s \le R \quad \text{implies that} \quad u(x) \in \overline{G}_2$$

and

$$(2.9)\quad (b)\qquad \|u_0 - u_0^k\|_s \le C\,\frac{R}{4}\ , \qquad k = 0,1,2,3,\ldots,$$

where $C \le 1$ is the constant from (1.16)(a) corresponding to G_2, i.e.,

$$(2.10)\qquad CI \le A_0(u) \le C^{-1}\,I \qquad \text{for al}\ \ u \in \overline{G}_2 \ .$$

With the facts in (2.9) and (2.10), it is evident by induction that u^{k+1} is well defined on the time interval $[0,T_k]$ and in fact belongs to $C^\infty([0,T_k] \times R^N)$ where $T_k > 0$ is the largest time where the estimate

(2.11) $$|||u^k - u_0^0|||_{s,T_k} \le R$$

is valid. Of course, as regards the nonlinear problem in (2.6), it is especially important that the following estimate is true:

(2.12) There is a $T_* > 0$ so that T_k from (2.11) satisfies $T_k \ge T_*$, $k = 1,2,3,\ldots$.

This is achieved through the following crucial lemma (which is proved by induction since $T_0 = +\infty$).

Lemma 2.1. (Boundedness in the High Norm) There are constants $L > 0$, $T_* > 0$ so that the solutions $u^{k+1}(x,t)$ defined in (2.7)(b) for $k = 0,1,2,3,\ldots$ satisfy

(A) $$|||u^{k+1} - u_0^0|||_{s,T_*} \le R$$

(B) $$|||\frac{\partial u^{k+1}}{\partial t}|||_{s-1,T_*} \le L .$$

We assume Lemma 2.1 for the moment and continue the argument. The next step in the argument is to find an appropriate norm so that

$$\sum_{k=0}^{\infty} \|u^{k+1} - u^k\| < \infty ,$$

i.e., to guarantee that there exists a u so that

(2.13) $$u = \lim_{k \to \infty} u^k.$$

Formally, if the norm $\|\cdot\|$ is strong enough, then as a consequence of (2.13) as $k \to \infty$,

(2.14)
$$A_j(u^k) \to A_j(u) , \qquad 0 \le j \le N$$
$$\frac{\partial u^k}{\partial x_j} \to \frac{\partial u}{\partial x_j} , \qquad 0 \le j \le N$$
$$u_0^k \to u_0 .$$

Thus, from (2.7)(b) we obtain the desired nonlinear solution of (2.6) by taking the limit. Which choice do we use for the strong norm $\|\cdot\|$? The first guess of $|||\cdot|||_{s,T_{**}}$, $T_{**} < T_*$ has severe technical problems. For example, we compute that

$$(2.15)\quad (a)\qquad A_0(u^k)\,\frac{\partial(u^{k+1}-u^k)}{\partial t} + \sum_{j=1}^{N} A_j(u^k)\,\frac{\partial}{\partial x_j}(u^{k+1}-u^k) = F_k$$

where

$$(2.15)\quad (b)\qquad F_k = \sum_{j=0}^{N} (A_j(u^k) - A_j(u^{k-1}))\,\frac{\partial u^k}{\partial x_j}\ .$$

If we naively apply the energy estimates from Section 1 to the differentiated equations, we can estimate

$$\|u^{k+1} - u^k\|_{s,T_*} \leq \alpha_k \|u^k - u^{k-1}\|_{s,T_*}\ ,$$

but α_k *depends* on $\|u^k\|_{s+1}$ *which is not bounded!* The idea of Kato and Lax to avoid this difficulty is that it is sufficient to prove in the low norm that

$$\sum_{k=0}^{\infty} \|u^{k+1} - u^k\|_{0,T_{**}} < \infty$$

to guarantee both (2.13) and (2.14). Thus we have the following lemma:

Lemma 2.2. (Contraction in the Low Norm) There exists $T_{**} \leq T_*$, $\alpha < 1$, and $\{\beta_k\}_{k=1}^{\infty}$ with $\Sigma|\beta_k| < \infty$ so that the functions u^{k+1} computed in (2.7)(b) satisfy

$$(2.16)\qquad |||u^{k+1} - u^k|||_{0,T_{**}} \leq \alpha |||u^k - u^{k-1}|||_{0,T_{**}} + \beta_k\ , \qquad k = 1,2,3,\ldots\ .$$

Given Lemma 2.1, the proof of Lemma 2.2 is very easy. We use equation (2.15)(a) and apply the energy estimate described in (1.21) to obtain

$$(2.17)\qquad |||u^{k+1} - u^k|||_{0,T} \leq \hat{C}\, e^{\hat{C}T}(\|u_0^{k+1} - u_0^k\|_0 + T|||F_k|||_{0,T})\ .$$

From (2.4)(c) it follows that

$$\|u_0^{k+1} - u_0^k\| \leq \hat{C}\, 2^{-k}\ ,$$

and from the explicit form of F_k in (2.15)(b) we deduce from Lemma 2.1 and Taylor's theorem that

$$|||F_k|||_{0,T} \leq \hat{C}|||u^k - u^{k-1}|||_{0,T}\,.$$

From these two facts and the estimate in (2.17), we prove Lemma 2.2 with a suitable choice of T_{**} and $\beta_k = \hat{C}\, 2^{-k}$.

Using Lemma 2.1 and Lemma 2.2, we complete the proof of Theorem 2.1(a). Our starting point is the following elementary remark:

If $\|u^{k+1} - u^k\| \leq \alpha\|u^k - u^{k-1}\| + \beta_k$ and $\alpha < 1$, $\sum_{k=1}^{\infty} |\beta_k| < \infty$,

then $\sum_{k=1}^{\infty} \|u^{k+1} - u^k\| < \infty$.

From the above statement and Lemma 2.2, we conclude immediately that there exists $u \in C([0,T_{**}], L^2(R^N))$ so that

$$\lim_{k\to\infty} |||u^k - u|||_{0,T_{**}} = 0 . \tag{2.18}$$

Furthermore, from Lemma 2.1, we have

$$\text{(2.19)}\quad \begin{aligned} &\text{(a)} \quad |||u^k|||_{s,T_{**}} + |||\frac{\partial u^k}{\partial t}|||_{s-1,T_{**}} \leq \hat{C} \\ &\text{(b)} \quad u^k(x,t) \in \overline{G}_2 \subset\subset G , \qquad (x,t) \in R^N \times [0,T_{**}] . \end{aligned}$$

Also, the Sobolev space interpolation inequalities imply that for any s' with $0 \leq s' \leq s$,

$$\|v\|_{s'} \leq C_s \|v\|_0^{1-s'/s} \|v\|_s^{s'/s} \tag{2.20}$$

so that (2.19)(a) and (2.20) imply for any $s' < s$

$$|||u^k - u^\ell|||_{s',T_{**}} \leq \hat{C}(|||u^k - u^\ell|||_{0,T_{**}})^{1-s'/s} . \tag{2.21}$$

(For this argument we will use well-known properties of the spaces $H^{s'}(R^N)$ for s' any positive number.) From (2.18) and (2.21) we conclude that

$$\lim_{k\to\infty} |||u^k - u|||_{s',T_{**}} \to 0 , \qquad \text{for any } s' < s , \tag{2.22}$$

so that if we choose $s' > \frac{n}{2} + 1$, Sobolev's lemma implies

$$u^k \to u \quad \text{in} \quad C([0,T_{**}],C^1(R^N)) . \tag{2.23}$$

From (2.23) and the formula in (2.7)(b),

$$\frac{\partial u^{k+1}}{\partial t} = - A_0^{-1}(u^k) \sum_{j=1}^{N} A_j(u^k) \frac{\partial u^{k+1}}{\partial x_j}$$

so we see that

$$\frac{\partial u^k}{\partial t} \to \frac{\partial u}{\partial t} \quad \text{in} \quad C([0,T_{**}],C(R^N))$$

and also that u belongs to $C^1([0,T_*]\times R^N)$ and is a classical solution of (2.6). The additional fact that u belongs to $C_w([0,T_{**}],H^s)\cap \mathrm{Lip}([0,T_{**}],H^{s-1})$ follows directly from the estimate in (2.19)(a) and the strong convergence in $C([0,T_{**}],H^{s'})$ established in (2.22) combined with an elementary argument. For example, to prove that $u(t)\in C_w([0,T_{**}],H^s)$, we let $[\phi,u]$ for $\phi\in H^{-s}$, $u\in H^s$ denote the dual pairing of H^{-s} and H^s through the L^2 inner product. Since from (2.22) $u^k\to u$ in $C([0,T],H^{s'})$ for $s'<s$, it follows that

$$[\phi,u^k(t)]\to[\phi,u(t)]$$

uniformly on $[0,T]$ for any $\phi\in H^{-s'}$. Now,

$$|||u_k|||_{s,T_{**}} \le R + \|u_0^0\|_s$$

and $H^{-s'}$ is dense in H^{-s} for $s'<s$ so by an $\varepsilon/3$-argument using the above bound,

(2.24) $$[\tilde{\phi},u^k(t)]\to[\tilde{\phi},u(t)] \qquad \text{as} \quad k\to\infty$$

uniformly on $[0,T_{**}]$ for any $\tilde{\phi}\in H^{-s}$ - this fact implies that $u(t)\in C_w([0,T_{**}],H^s)$ since the uniform limit of continuous functions is continuous. This completes the proof of Theorem 2.1(a).

It remains for us to discuss the proof of Lemma 2.1. We set $v^{k+1}=u^{k+1}-u_0^0$ and compute that v^{k+1} satisfies

(2.25) (a)
$$A_0(u^k)\frac{\partial v^{k+1}}{\partial t}+\sum_{j=1}^{N}A_j(u^k)\frac{\partial v^{k+1}}{\partial x_j}=H^k$$
$$v^{k+1}(x.0)=u_0^{k+1}(x)-u_0^0(x)$$

where

(2.25) (b) $$H^k=\sum_{j=1}^{N}A_j(u^k)\frac{\partial}{\partial x_j}u_0^0 .$$

Also, by the induction hypothesis, $u^k(x,t)\in\overline{G}_2$, $(x,t)\in R^N\times[0,T_*]$, where T_* will be chosen later through higher order a priori energy estimates.

For further reference, we ignore the superscripts k and consider $u\in C^\infty$, $v\in C^\infty$, $u(x,t)\in\overline{G}_2$, $(x,t)\in R^N\times[0,T_*]$ satisfying

$$A_0(u)\frac{\partial v}{\partial t} + \sum_{j=1}^{N} A_j(u)\frac{\partial v}{\partial x_j} = H \tag{2.26}$$
$$v(x,0) = v_0(x) .$$

We define v_α by $v_\alpha = D^\alpha v$ for $|\alpha| \le s$. If we divide the equation in (2.26) by $A_0(u)$ and differentiate with respect to x, α-times, we compute the equations, for $|\alpha| \le s$,

$$A_0(u)\frac{\partial v_\alpha}{\partial t} + \sum_{j=1}^{N} A_j(u)\frac{\partial v_\alpha}{\partial x_j} = A_0\, D^\alpha(A_0^{-1}\, H) + F_\alpha \tag{2.27}$$
$$v_\alpha(x,0) = D^\alpha v_0(x) ,$$

with F_α defined by the commutator terms as

$$F_\alpha = \sum_{j=1}^{N} A_0(u)[(A_0^{-1}A_j)(u)\frac{\partial v_\alpha}{\partial x_j} - D^\alpha(A_0^{-1}A_j\frac{\partial v}{\partial x_j})] , \qquad |\alpha| \le s . \tag{2.28}$$

The strategy for the proof of Lemma 2.1 is now evident. We assume that u^k satisfies (A), (B) for appropriate choices of L, T_* and use the differential equations and the energy principle in (1.21) to estimate $|||u^{k+1} - u_0^0|||_{s,T_*}$; from (2.7)(b) it is clear (by solving for $\frac{\partial u^{k+1}}{\partial t}$) that the fixed number R in Lemma 2.1 (A) *automatically* determines an a priori choice for L in (B) of Lemma 2.1 independent of T_* (one can apply Proposition 2.1 formulated below). In estimating $|||u^{k+1} - u_0^0|||_{s,T}$ by the energy principle, we need to control the inhomogeneous terms on the right-hand side of (2.27) for $|\alpha| \le s$ with $u = u^k$, $v = v^{k+1}$, and H^k from (2.25)(b). In particular, it is crucial that we find a constant $\tilde{C}$, depending only on $\bar{G}_2$, $\|u\|_s$, R, and s, i.e., $\tilde{C}(\bar{G}_2, \|u\|_s, R, s)$ so that

$$\Big(\sum_{1\le|\alpha|\le s} \|F_\alpha\|_0^2\Big) + \Big(\sum_{|\alpha|\le s} \|A_0 D^\alpha(A^{-1}H)\|_0^2\Big) \le \tilde{C}(\bar{G}_2, \|u\|_s, R, s)(\|v\|_s^2 + 1) . \tag{2.29}$$

We assume (2.29) for the moment and continue the proof. By applying the energy principle above (1.21) to v_α^{k+1} from (2.27), using (2.29), and summing over α with $|\alpha| \le s$, we obtain from (1.21) that

$$|||u^{k+1} - u_0^0|||_{s,T} \le C^{-1}\, e^{(L\tilde{C} + \tilde{C})T}(\|u_0^{k+1} - u_0^0\|_s + T\tilde{C})$$

with C the constant from (2.10). From (2.9)(a) and (2.5), we have

$$\|u_0^{k+1} - u_0^0\|_s \le C\,\frac{R}{2}$$

so that the above two inequalities imply

$$|||u^{k+1} - u_0^0|||_{s,T} \le e^{(L+1)\tilde{C}T}(\frac{R}{2} + T\tilde{C}) .$$

The fixed a priori choice of $T_* > 0$ guaranteeing the inductive step in Lemma 2.1(A) is now easily determined from the above inequality. This completes the proof of Lemma 2.1 provided that we can verify the estimate in (2.29).

We won't prove the estimate in (2.29) in detail here since we give similar calculations in complete detail when we discuss the continuation principle in Theorem 2.2 below. The key idea in the proof of (2.29) is to apply calculus inequalities in Sobolev spaces - the simplest of these inequalities is the following fact:

(2.30) if $u,v \in H^s$, $s > \frac{N}{2}$, then $u \cdot v \in H^s$ and $\|uv\|_s \le C_s\|u\|_s\|v\|_s$.

Here we will state three rather sharp calculus inequalities of the type first formulated by Moser [24], which give a precise version of the estimates stated in (2.29) -we will give several applications of these inequalities in subsequent sections of these lectures.

Proposition 2.1. (Moser-type Calculus Inequalities)

(A) For $f,g \in H^s \cap L^\infty$ and $|\alpha| \le s$

$$\|D^\alpha(fg)\|_0 \le C_s(|f|_{L^\infty}\|D^s g\|_0 + |g|_{L^\infty}\|D^s f\|_0) .$$

(B) For $f \in H^s$, $Df \in L^\infty$, $g \in H^{s-1} \cap L^\infty$ and $|\alpha| \le s$

$$\|D^\alpha(fg) - fD^\alpha(g)\|_0 \le C_s(|Df|_{L^\infty}\|D^{s-1}g\|_0 + |g|_{L^\infty}\|D^s f\|_0) .$$

(C) Assume $g(u)$ is a smooth vector-valued function on G, $u(x)$ is a continuous function with $u(x) \in G_1$, $\overline{G}_1 \subset\subset G$, and $u(x) \in L^\infty \cap H^s$. Then for $s \ge 1$,

$$\|D^s g(u)\|_0 \le C_s \left|\frac{\partial g}{\partial u}\right|_{s-1,\overline{G}_1} |u|_{L^\infty}^{s-1}\|D^s u\|_0 .$$

Here $|\cdot|_{r,\overline{G}_1}$ is the C^r norm on the set $\overline{G}_1$. We remark that part (A) of Proposition 2.1 already implies a rather sharp version of the Banach algebra properties stated in (2.30). The proofs of (A), (B), and (C) all are based upon the celebrated inequality of Gagliardo-Nirenberg,

(2.31) For $u \in H^s \cap L^\infty$, $|D^i u|_{L^{2s/i}} \le C_s|u|_{L^\infty}^{1-(i/s)}\|D^s u\|_0^{i/s}$, together with repeated applications of Hölder's inequality. Here $|u|_{L^p}$ for $1 \leq p \leq +\infty$ is defined by

$$|u|^p_{L^p} = \int_{R^N} |u|^p .$$

Detailed proofs for (A), (B), and (C) can be found in [24] and the appendix of [15], so we won't prove this proposition here - variants of Proposition 2.1 are also quite useful (see appendix of [15]).

Remark 1. Below we often apply (C) with $g(u) = G(u+u_0)$ with u_0 a fixed constant vector and G a smooth function. The next remark on the proof is useful when we prove Theorem 2.1(b) in the next section.

Remark 2. Let $u(x,t)$ be the local solution from Theorem 2.1(a). As a consequence of (2.19)(b) and (2.22), we conclude that

$$\text{(2.31)} \quad \begin{aligned} &\text{(a)} \quad A_0(u(x,t)) \in C([0,T_*], C(R^N)) \\ &\text{(b)} \quad CI \leq A_0(u(x,t)) \leq C^{-1}I, \qquad (x,t) \in R^N \times [0,T_{**}] . \end{aligned}$$

We define the norm $\|v\|^2_{s,A_0(t)}$ by

$$\text{(2.32)} \qquad \|v\|^2_{s,A_0(t)} = \sum_{|\alpha|\leq s} \int_{R^N} (D^\alpha v, A_0(t) D^\alpha v)dx$$

for $0 \leq t \leq T_{**}$ with $A_0(t)$, the short-hand notation for $A_0(u(x,t))$. From (2.31) we have the two simple facts that

$$\text{(2.33)} \quad \begin{aligned} &\text{(a)} \quad C\|v\|^2_s \leq \|v\|^2_{s,A_0(t)} \leq C^{-1}\|v\|^2_s \\ &\text{(b)} \quad \overline{\lim_{t\downarrow 0}} \|v(t)\|^2_{s,A_0(t)} = \overline{\lim_{t\uparrow 0}} \|v(t)\|^2_{s,A_0(0)} \end{aligned}$$

for any $v(t) \in C_w([0,T_{**}], H^s)$.

The Proof of Theorem 2.1(b)

We begin the proof with several easy reductions. First, it is sufficient to prove that $u \in C([0,T], H^s)$ since it follows directly from the equation in (2.6) that $u \in C^1([0,T], H^{s-1})$. Furthermore, we only need to prove the strong right continuity of u at $t = 0$ since the same argument given below will prove the strong right continuity at any other $\hat{T}$ with $0 \leq \hat{T} < T$; furthermore, the equation in (2.6) is *reversible in time* and the argument we give below is time reversible too

- thus, the proof of strong right continuity on $[0,T)$ implies the strong left continuity on $(0,T]$ and therefore $u \in C([0,T],H^s)$.

Thus, our task is reduced to verifying the strong right continuity of u at $t = 0$. We consider H^s with the equivalent norm $\|\cdot\|^2_{s,A_0(0)}$ defined in (2.32). Now, if $\{w_n\}$ converges weakly in a Hilbert space to w, w_n converges strongly to w if and only if $\|w\| \geq \overline{\lim}_{n\to\infty} \|w_n\|$. By applying this elementary fact in the Hilbert space H^s with norm $\|\cdot\|^2_{s,A_0(0)}$ and (2.33), we will prove the strong right continuity of u at zero in H^s provided that we establish

$$(2.34)\qquad \|u_0\|^2_{s,A_0(0)} \geq \overline{\lim_{t\downarrow 0}} \|u(t)\|^2_{s,A_0(0)} = \overline{\lim_{t\downarrow 0}} \|u(t)\|^2_{s,A_0(t)} .$$

The fact in (2.34) is an immediate consequence of the following.

Lemma 2.3. If u is the local solution of Theorem 2.1(a) on some interval $[0,T_{**}]$, then there is $f(s) \in L^1([0,T_{**}])$ so that

$$(2.35)\qquad \begin{aligned} \|u(t)\|^2_{s,A_0(t)} &= \sum_{|\alpha|\leq s} \int_{R^N} (D^\alpha u, A(u(t))D^\alpha u)\,dx \\ &\leq \sum_{|\alpha|\leq s} \int_{R^N} (D^\alpha u_0, A(u_0)D^\alpha u_0) + \int_0^t |f(s)|\,ds \\ &= \|u_0\|^2_{s,A_0(0)} + \int_0^t |f(s)|\,ds , \qquad 0 < t \leq T_{**} . \end{aligned}$$

We only need to take the lim sup as $t \downarrow 0$ of both sides of (2.35) to deduce the fact in (2.34) required to complete the proof of Theorem 2.1(b).

It remains for us to prove (2.35). We prove the estimate by examining the iteration scheme, using the energy principle in (1.20) (not the consequence in (1.21)), and letting $k \to \infty$. Recall that the functions u^k are in $C^\infty \cap H^s$. By the energy principle for smooth solutions, for $0 \leq t \leq T_{**}$,

$$(2.36)\qquad \frac{\partial}{\partial t} \sum_{|\alpha|\leq s} \int_{R^N} (D^\alpha u^{k+1}, A_0(u^k)D^\alpha u^{k+1}) = \int_{R^N} (\operatorname{div} \vec{A}(u^k)u^{k+1}, u^{k+1}) + 2\int_{R^N} (F_s^{k+1}, u^{k+1}) .$$

With $u_\alpha^{k+1} = D^\alpha u^{k+1}$, F_s^{k+1} is given by

$$F_s^{k+1} = \sum_{\substack{1\leq|\alpha|\leq s\\ 1\leq j\leq N}} A_0(u^k)\left[A_0(u^k)^{-1} A_j(u^k) \frac{\partial u_\alpha^{k+1}}{\partial x_j} - D^\alpha (A_0^{-1} A_j(u^k) \frac{\partial u^{k+1}}{\partial x_j}\right] .$$

As a consequence of the bounds in the high norm in Lemma 2.1 for u^k and Proposition 2.1, it is clear that the right-hand side of (2.36) is bounded by an integrable scalar function $f(s)$ - in fact, $f(s) \in L^\infty([0,T_{**}])$ (this could be exploited further to weaken hypotheses, but we won't do this here). Thus, after integrating (2.36), we have for $0 \leq t \leq T_{**}$

$$(2.37) \quad \sum_{|\alpha|\leq s} \int_{R^N} (D^\alpha u^{k+1}(t), A_0(u^k(t)) D^\alpha u^{k+1}(t)) \leq \sum_{|\alpha|\leq s} \int_{R^N} (D^\alpha u_0^{k+1}, A_0(u_0^k) D^\alpha u_0^{k+1}) + \int_0^t |f(s)| ds .$$

Now, $\{u_0^k\}$ is defined in (2.5) through mollification, so it follows from the properties in (2.4) that

$$\lim_{k\to\infty} \sum_{|\alpha|\leq s} \int_{R^N} (D^\alpha u_0^{k+1}, A_0(u_0^k) D^\alpha u_0^{k+1}) = \|u_0\|^2_{s,A_0(0)} .$$

Also, from (2.22), as $k \to \infty$,

$$\max_{0\leq t\leq T_{**}} |A_0(u^k(t)) - A_0(u(t))|_{L^\infty} \to 0 ,$$

and furthermore, from the weak convergence as $k \to \infty$ in (2.24) that

$$\overline{\lim_{k\to\infty}} (u^{k+1}(t), u^{k+1}(t))_{s,A_0(t)} \geq (u(t), u(t))_{s,A_0(t)} .$$

These two facts and (2.33) imply

$$\|u(t)\|^2_{s,A(t)} \leq \overline{\lim_{k\to\infty}} \sum_{|\alpha|\leq s} \int (D^\alpha u^{k+1}, A_0(u^k) D^\alpha u^{k+1}) .$$

Now (2.35) follows from (2.37) and the additional facts proved above. This completes the lemma and the proof of Theorem 2.1(b).

2.1. *A Continuation Principle for Smooth Solutions*

Our objective here is to prove the following more precise version of Theorem 2.2 stated in the introduction:

Theorem 2.2. (A Sharp Continuation Principle) Assume that $u_0 \in H^s$ for some $s > \frac{N}{2} + 1$. Let $T > 0$ be some given time. Assume that there are fixed constants M_1, M_2 and a fixed open set G_1 with $\overline{G}_1 \subset\subset G$ (all independent of T_*) so that for any interval of classical existence $[0,T_*]$, $T_* \leq T$ for $u(t)$ from Theorem 2.1, the following a priori estimates are satisfied:

1) $\quad |\mathrm{div}\,\vec{A}|_{L^\infty} \leq M_1 , \qquad 0 \leq t \leq T_*$

2) $|Du|_{L^\infty} \le M_2$, $\qquad 0 \le t \le T_*$

3) $u(x,t) \in \overline{G}_1 \subset\subset G$, $\qquad (x,t) \in R^N \times [0,T_*]$.

Then the classical solution $u(t)$ exists on the interval $[0,T]$, with $u(t)$ $C([0,T],H^s) \cap C^1([0,T],H^{s-1})$. Furthermore, $u(t)$ satisfies the a priori estimate

$$\|\!|u|\!\|_{s,T_*} \le C \exp((M_1 + M_2)CT_*)\|u_0\|_s \tag{2.38}$$

for T_* with $0 \le T_* \le T$ and the two constants C in (2.38) depend only on s and $\overline{G}_1$, i.e., $C(s,\overline{G}_1)$.

Theorem 2.2 has two immediate corollaries. (We leave the easy proofs using the local existence Theorem 2.1 and Theorem 2.2 as an exercise for the reader.)

Corollary 1. Assume $u_0 \in H^s$ for some $s > \frac{N}{2} + 1$. Assume that $u(x,t)$ is a classical solution of (2.1) on some interval $[0,T]$ with $u \in C^1([0,T] \times R^N)$, then on the same time interval $[0,T]$ necessarily $u \in C([0,T],H^s)$. In particular, if $u_0 \in \bigcap_s H^s$, on any interval $[0,T]$ where u belongs to $C([0,T],H^{s_0})$ for some s_0, $s_0 > (N/2) + 1$, automatically u is a function in $C^\infty([0,T] \times R^N)$.

Corollary 2. Assume $u_0 \in H^s$ for some $s > \frac{N}{2} + 1$. Then $[0,T)$ with $T < \infty$ is a maximal interval of H^s existence if and only if either

(1) $|u_t|_{L^\infty} + |Du|_{L^\infty} \to \infty$ $\qquad$ as $t \uparrow T$,

or

(2) as $t \uparrow T$, $u(x,t)$ escapes every compact subset $K \subset\subset G$.

Next, we prove Theorem 2.2. It is sufficient to establish the a priori estimate in (2.38) on any interval $[0,T_*]$, $T_* \le T$ of local H^s existence guaranteed by Theorem 2.1 by the following standard reasoning: if $[0,T_*)$ with $T_* \le T$ was a maximal interval of H^s local existence, we could apply the a priori estimate from (2.38) and the local existence theorem, Theorem 2.1, beginning at the time $T_* - \varepsilon$, for any $\varepsilon > 0$ to continue this solution beyond T_* – a contradiction. To prove the a priori estimate in (2.38) we utilize the sharp calculus inequalities given in Proposition 2.1. With the notation from (2.6), u satisfies the symmetric hyperbolic system

$$\begin{aligned} A_0(u)u_t + \sum_{j=1}^{N} A_j(u)u_{x_j} &= 0 \\ u(x,0) &= u_0(x) \end{aligned} \tag{2.39}$$

where as a consequence of the hypothesis in (3) of Theorem 2.2,

$$CI \leq A_0(u(x,t)) \leq C^{-1} I . \tag{2.40}$$

Next, we compute as in (2.27) that $u_\alpha = D^\alpha u$ with $|\alpha| \leq s$ satisfies the equation

$$A_0(u) \frac{\partial u_\alpha}{\partial t} + \sum_{j=1}^{N} A_j(u) \frac{\partial u_\alpha}{\partial x_j} = F_\alpha , \tag{2.41}$$

with

$$F_\alpha = \sum_{n=1}^{N} A_0(u) [D^\alpha (A_0^{-1} A_j(u) \frac{\partial u}{\partial x_j}) - A_0^{-1} A_j(u) \frac{\partial u_\alpha}{\partial x_j}] . \tag{2.42}$$

We estimate F_α for $1 \leq |\alpha| \leq s$ by utilizing Proposition 2.1. By (2.40) and (B) of Proposition 2.1, we obtain

$$\begin{aligned} \sum_{1\leq|\alpha|\leq s} \|F_\alpha\|_0 &\leq \sum_{\substack{1\leq|\alpha|\leq s \\ 1\leq j\leq N}} C^{-1} \|D^\alpha (A_0^{-1} A_j \frac{\partial u}{\partial x_j}) - A_0^{-1} A_j \frac{\partial u_\alpha}{\partial x_j}\|_0 \\ &\leq C_s \sum_{j=1}^{N} \left(|D(A_0^{-1} A_j)|_{L^\infty} \|D^{s-1} \frac{\partial u}{\partial x_j}\|_0 + |\frac{\partial u}{\partial x_j}|_{L^\infty} \|D^s (A_0^{-1} A_j(u))\|_0 \right) . \end{aligned} \tag{2.43}$$

Next, we apply (C) of Proposition 2.1 to get

$$\|D^s (A_0^{-1} A_j(u)\|_0 \leq C(s, \overline{G}_1) \|D^s u\|_0 \tag{2.44}$$

so that (2.43) and (2.44) imply the bound

$$\sum_{1\leq|\alpha|\leq s} \|F_\alpha\|_0 \leq C(s, \overline{G}_1) M_2 \|D^s u\|_0 \tag{2.45}$$

where we have used the estimate in 2) of Theorem 2.2. Finally, we compute

$$\frac{\partial}{\partial t} \sum_{0\leq|\alpha|\leq s} \int_{R^N} (u_\alpha, A_0(u) u_\alpha)$$

by the energy principle from (1.20) and use the sharp estimate in (2.45), (2.40), and Gronwall's inequality (in standard fashion) as in (1.21) to prove the a priori inequality in (2.38), as required to finish the proof. We remark that the a priori bound in 1) of Theorem 2.2 is needed in this last step - see (1.20). Also, a sharp version of the estimate in (2.29) can be proved in exactly the same fashion as we did above in (2.43) - (2.45).

2.2. *Uniformly Local Sobolev Spaces*

In the beginning of Section 2 and in Subsection 2.1 we have given a complete treatment of the H^s classical existence theory under the assumption that $u_0(x) \in H^s(R^N)$; these results apply to the physical examples discussed in Section 1 provided that the initial data differs from a constant state u_0 by an H^s function. However, such conditions on the initial data are not always natural since for example the initial density might be a smooth plane wave $\rho_0(x \cdot \omega)$ with $\lim_{\tilde{x} \to +\infty} \rho_0(\tilde{x}) = \rho_+$ and $\lim_{\tilde{x} \to -\infty} \rho_0(\tilde{x}) = \rho_-$ with $\rho_-, \rho_+ > 0$ and $\rho_- \neq \rho_+$. The proofs in the previous two subsections were based upon the global energy principle for symmetric hyperbolic systems described in Section 1. However, it is well known that solutions of hyperbolic equations have finite propagation speed and obey a local energy principle.

To take advantage of the above local energy principle and also to allow for initial data like the density in the previous paragraph, Kato [14] introduced the uniformly local Sobolev spaces $H^s_{u\ell}$. These spaces are defined in the following fashion: Let $\theta \in C_0^\infty(R^N)$ be a function so that $\theta \geq 0$ and

$$\theta(x) = \begin{cases} 1, & |x| \leq \frac{1}{2} \\ 0, & |x| > 1 \end{cases} \tag{2.46}$$

and define $\theta_{d,y}(x)$ by

$$\theta_{d,y}(x) = \theta\left(\frac{x-y}{d}\right). \tag{2.47}$$

Definition. The function u belongs to the *uniformly local Sobolev spaces* $u \in H^s_{u\ell}$, provided that there is some $d > 0$ so that

$$\max_{y \in R^N} \|\theta_{d,y} u\|_s = \tilde{\|} u \tilde{\|}_{s,d} < \infty. \tag{2.48}$$

The norms $\tilde{\|} \cdot \tilde{\|}_{s,d}$ are all equivalent norms on $H^s_{u\ell}$ as d varies; in particular,

$$\tilde{\|} u \tilde{\|}_{s,d_1} \leq C \tilde{\|} u \tilde{\|}_{s,d_2} \tag{2.49}$$

provided that d_1, d_2 satisfy $0 < d_- \leq d_1, d_2 \leq d_+$.

Next, we briefly formulate the local energy principle for the linear symmetric hyperbolic equation

$$\begin{aligned} A_0(u)v_t + \sum_{j=1}^N A_j(u)v_{x_j} &= F \\ v(x,0) &= v_0(x) \end{aligned} \tag{2.50}$$

where $u(x,t) \in G_1$, $\overline{G}_1 \subset\subset G$. We recall that

$$\text{(2.51)} \quad \text{(a)} \qquad CI \le A_0(u) \le C^{-1}I \qquad \text{for } u \in \overline{G}_1$$

and also there is a number $D > 0$ so that

$$\text{(2.51)} \quad \text{(b)} \qquad \max_{\substack{|\omega|=1 \\ u \in \overline{G}_1}} \Big| \Big(\sum_{j=1}^{N} A_j(u)\omega_j v, v \Big) \Big| \le D|v|^2 .$$

Given (2.51), we define the number R by

$$\text{(2.52)} \qquad R = \frac{D}{2C} .$$

By using Green's formula as in (1.19), (1.20), we obtain the *local energy principle*,

$$\text{(2.53)} \qquad \int_{|x-y|\le d} dx\,(A_0(u)v,v)(T) \le \int_0^T \int_{|x-y|\le d+R(T-s)} [2(F,v) + ((\operatorname{div} \vec{A})v,v)]\,dx\,ds + \int_{|x-y|\le d+RT} (A_0(u)v_0,v_0)\,dx .$$

This energy principle and the estimate analogous to (1.21) obtained by Gronwall's inequality are two of the main ingredients needed in generalizing Theorems 2.1, 2.2 to initial data in $H^s_{u\ell}$. In particular, we remark that when $F = 0$ in (2.50), we conclude from (2.51) - (2.53), the energy estimate

$$\text{(2.54)} \qquad \tilde{\|} v \tilde{\|}_{0,d}(T) \le C^{-1} \exp(C^{-1}|\operatorname{div} \vec{A}|_{L^\infty} T) \tilde{\|} v_0 \tilde{\|}_{0,2d+RT}$$

which illustrates a typical generalization of (1.21) using $H^s_{u\ell}$ norms.

The other technical ingredient is the following analogue of the calculus inequalities in Proposition 2.1:

Proposition 2.2. (Calculus Inequalities for $H^s_{u\ell}$*)*

(A) For $f,g \in H^s_{u\ell} \cap L^\infty$ and $|\alpha| \le s$

$$\tilde{\|} D^\alpha(fg) \tilde{\|}_{0,d} \le C_{s,d}(|f|_{L^\infty} \tilde{\|} g \tilde{\|}_{s,2d} + |g|_{L^\infty} \tilde{\|} f \tilde{\|}_{s,2d}) .$$

(B) For $f \in H^s_{u\ell} \cap C^1$, $g \in H^{s-1}_{u\ell} \cap L^\infty$ and $|\alpha| \le s$

$$\tilde{\|} D^\alpha(fg) - fD^\alpha g \tilde{\|}_{0,d} \le C_{s,d}(|f|_{C^1} \tilde{\|} g \tilde{\|}_{s-1,2d} + |g|_{L^\infty} \tilde{\|} f \tilde{\|}_{s,2d}) .$$

(C) For $g(u)$ smooth on G, $u(x)$ continuous with $u(x) \in \overline{G}_1 \subset\subset G$, and $u \in H^s_{u\ell}$

$$\tilde{\|} g(u) \tilde{\|}_{s,d} \leq C(\overline{G}_1, s, d)(1 + \tilde{\|} u \tilde{\|}_{s,2d}) .$$

To prove Proposition 2.2, one only needs to repeat the proofs in Proposition 2.1 appropriately inserting and keeping track of the cut-offs $\theta_{y,d}$. With the facts in (2.53), (2.54), and Proposition 2.2, it is evident that we can repeat the proofs of Theorems 2.1, 2.2 to get analogous results when the initial data u_0 belongs to $H^s_{u\ell}$.

We state these results but omit any discussion of the proofs, leaving them as an exercise for the interested reader. (The corollaries to Theorem 2.2 also have obvious analogues which we don't state here explicitly.)

Theorem 2.1. (Annex) Assume $u_0 \in H^s_{u\ell}$, $s > \frac{N}{2} + 1$ and $u_0(x) \in G_1$, $\overline{G}_1 \subset\subset G$. Then there is a time interval $[0,T]$ with $T > 0$ so that the equations in (2.1) have a unique classical solution $u \in C^1([0,T] \times R^N)$ with $u(x,t) \in G_2$, $\overline{G}_2 \subset\subset G$ for $(x,t) \in R^N \times [0,T]$. Furthermore, $u(t) \in C([0,T], H^s_{loc}) \cap C^1([0,T], H^{s-1}_{log})$ and $u(t) \in L^\infty([0,T], H^s_{u\ell})$; also T depends only on s and $\tilde{\|} u_0 \tilde{\|}_{s,d}$.

Theorem 2.2. (Annex) If $u_0 \in H^s_{u\ell}$, $s > \frac{N}{2} + 1$, and the same three a priori estimates in Theorem 2.2 are satisfied for any time interval $[0,T_*]$ of local $H^s_{u\ell}$ existence with $T_* \leq T$, then the classical $H^s_{u\ell}$ solution of (2.1) exists on the interval $[0,T]$ with $u(t) \in C([0,T], H^s_{loc}) \cap C^1([0,T], H^{s-1}_{loc}) \cap L^\infty([0,T], H^s_{u\ell})$. A similar a priori estimate like the one in (2.38) is also valid in $H^s_{u\ell}$ norms.

Some Special Results in 1-D Using the Maximum Norm

Finally, we end our general discussion of the smooth solution theory for (1.1) by mentioning some special results for hyperbolic systems in a single space dimension,

$$u_t + A(u)u_x = S(u,x,t)$$

$$u(x,0) = u_0(x)$$

with initial data $u_0(x)$ belonging to $C^1(R^1)$ with $|u_0|_{L^\infty} + |u_0'|_{L^\infty} < \infty$ and $u_0 \in \overline{G}_1 \subset\subset G$. The local existence of a $C^1([0,T] \times R^N)$ solution u to the above equations is true, but the proofs are surprisingly delicate. The results are due to Douglis [5] and Hartman and Winter [10] independently. Of course, the linearized equations are well posed in the *maximum* norm in a single space dimension - this is the reason that the hypothesis $u_0 \in H^s_{u\ell}$, $s > \frac{N}{2} + 1$, can be replaced by the weaker condition $u_0 \in C^1(R^1)$. A simpler version of the continuation principle in Theorem

2.2 is valid in the maximum norm. This can be deduced from an examination of the proof in [5] or [10], but we don't know a reference where this is done in detail.

3. Compressible and Incompressible Fluid Flow

A Formal Derivation of the Relationship Between Compressible and Incompressible Fluids

We begin this section by describing the formal asymptotic relationship between the equations of isentropic compressible fluid flow,

$$\frac{D\rho}{Dt} + \rho \operatorname{div} v = 0$$

$$\frac{Dv}{Dt} + \frac{1}{\rho}\nabla p = 0 \tag{3.1}$$

$$p = A\rho^r, \quad r > 1$$

and the incompressible Euler equations

$$\rho_0 \frac{Dv^\infty}{Dt} + \nabla p^\infty = 0$$

$$\operatorname{div} v^\infty = 0 . \tag{3.2}$$

We shall see below that the equations in (3.2) arise as the singular limiting equations for the system in (3.1) as the Mach number tends to zero. The first step in understanding this limiting process is through the nondimensionalization of the compressible fluid equations. We consider (3.1) with initial data

$$\rho(x,0) = \rho_0(x) , \qquad v(x,0) = v_0(x)$$

and set $\rho_m = \max \rho_0(x)$, $|v_m| = \max|v_0(x)|$ to represent typical values for the density and fluid velocity present in the initial data for (3.1). We introduce the new variables

$$\tilde{v} = \frac{v}{|v_m|} , \qquad \tilde{\rho} = \frac{\rho}{\rho_m}$$

$$x' = x , \qquad t' = |v_m|t$$

and rewrite the compressible fluid equations in (3.1). The result is that the equations of compressible fluid flow are written in the nondimensional form

$$\frac{D\tilde{\rho}}{Dt'} + \tilde{\rho} \operatorname{div} \tilde{v} = 0 , \qquad \tilde{\rho}\frac{D\tilde{v}}{Dt'} + \lambda^2 \nabla p(\tilde{\rho}) = 0 , \tag{3.3}$$

where λ^2 is the nondimensional quantity

$$\lambda^2 = \left(\frac{dp}{d\rho}(\rho_m)/|v_m|^2\right)(\gamma A)^{-1} .$$

In particular, the *Mach number* M is defined by the ratio of the *typical fluid speed* (described by $|v_m|$) to the typical sound speed (described by $(dp/d\rho)^{\frac{1}{2}}$), i.e., with $c(\rho_m) = (dp/d\rho)^{\frac{1}{2}}$,

(3.4) $$M = \frac{|v_m|}{c(\rho_m)} , \qquad \lambda = M^{-1}(\gamma A)^{-\frac{1}{2}} .$$

The main content of the formal derivation to be presented below is that the incompressible Euler equations from (3.2) have solutions which approximate solutions of the compressible Euler equations in (3.1) *provided that the Mach number is small*, i.e., the typical fluid velocity is much less than the speed of sound. For example, in air at room temperature we have

$$\gamma = 1.4 , \qquad c = 333 \text{ meters/sec}$$

and a typical fairly slow moving mean fluid velocity is given by $|v_m| = 10$ meters/sec so that (with $A \equiv 1$)

$$\lambda^2 = \frac{(33.3)^2}{1.4} \cong 650 .$$

Here λ^2 is rather large so that anticipating our results below, one can expect the equations in (3.2) to have solutions that approximate those in (3.1) under suitable initial conditions.

It is convenient to drop the primes and tildas and rewrite the equations in (3.3) in terms of (p^λ, v^λ) from (1.5) which satisfy

(3.5) (a) $$\begin{aligned} (\gamma p^\lambda)^{-1} \frac{Dp^\lambda}{Dt} + \operatorname{div} v^\lambda &= 0 \\ \rho(p^\lambda) \frac{Dv^\lambda}{Dt} + \lambda^2 \nabla p^\lambda &= 0 , \end{aligned}$$

with $\rho(p^\lambda)$ determined in (1.4) and the initial conditions

(3.5) (b) $$p^\lambda(x,0) = p_0^\lambda(x) , \qquad v^\lambda(x,0) = v_0^\lambda(x) .$$

Proceeding formally, we assume the asymptotic expansions

(3.6) $$\begin{aligned} p^\lambda &= p_0 + \lambda^{-1} p_1 + \lambda^{-2} p_2 + 0(\lambda^{-3}) \\ v^\lambda &= v^\infty + \lambda^{-1} v_1 + \lambda^{-2} v_2 + 0(\lambda^{-3}) . \end{aligned}$$

Before continuing the argument, it is convenient to recall the following familiar fact:

(3.7) Any $v \in L^2(R^N)$ has the unique orthogonal decomposition

$$v = w + \nabla\phi ,$$

where $\operatorname{div} w = 0$. We use the notation $w = Pv$, $\nabla\phi = (I-P)v$ and also below we will use the facts that

$$\|Pv\|_s \leq \|v\|_s$$

for any number $s \geq 0$ (see [26]).

We substitute the expansions in (3.6) into the velocity equations from (3.5)(a) and equate powers of λ to obtain

$$\text{(3.8)} \quad \begin{aligned} &\text{Order } \lambda^2 , && \nabla p_0 \equiv 0 \\ &\text{Order } \lambda , && \nabla p_1 \equiv 0 \\ &\text{Order zero,} && P(\rho(p_0))\frac{Dv^\infty}{Dt} \equiv 0 . \end{aligned}$$

In the last formula, we have applied P from (3.7) and used the fact that $P(\nabla p^\lambda) = 0$. In particular, (3.7) and (3.8) imply that there exists $P_0(t)$, a possibly time varying constant, and $p^\infty(x,t)$, a mean scalar pressure, so that

$$\text{(3.9)} \quad \begin{aligned} &\text{(a)} && \rho(p_0)\frac{Dv^\infty}{Dt} = -\nabla p^\infty \\ &\text{(b)} && p_0 \equiv P_0(t) . \end{aligned}$$

Next, we substitute the expansions in (3.6) into the pressure equation in (3.5)(a) and equate powers of λ. By using the information already determined in (3.9), we compute that the order zero terms in the pressure equation yield

$$\text{(3.10)} \quad (\gamma P_0(t))^{-1}\frac{dP_0(t)}{dt} + \operatorname{div} v^\infty = 0 .$$

Now, one of the terms in (3.10) is a scalar function of t alone while the other term involves, in general, a function of x and t; thus, consistency with (3.9) requires that each term separately vanish, i.e.,

$$\text{(3.11)} \quad \begin{aligned} &\operatorname{div} v^\infty = 0 \\ &P_0(t) \equiv P_0 , \qquad P_0 \text{ a constant.} \end{aligned}$$

Furthermore, the order λ terms in (3.8) indicate that we can set $p_1 \equiv 0$ by redefining P_0. Now, by summarizing the facts in (3.6), (3.9), and (3.11), we deduce that as the Mach number tends to zero, the solutions of the compressible fluid equations in (3.5) have the formal asymptotic expansion

$$
\begin{aligned}
p^\lambda &= P_0 + \lambda^{-2}(p^\infty(x,t)) + O(\lambda^{-3}) \\
v^\lambda &= v^\infty(x,t) + \lambda^{-1}(v_1(x,t)) + O(\lambda^{-2}) ,
\end{aligned}
\tag{3.12}
$$

where (p^∞, v^∞) satisfy the incompressible Euler equations

$$
\begin{aligned}
&\operatorname{div} v^\infty = 0 \\
&\rho_0 \frac{Dv^\infty}{Dt} = - \nabla p^\infty ,
\end{aligned}
\tag{3.13}
$$

with $\rho_0 = \rho(P_0)$. An obvious restriction on the initial data for the validity of the above expansion is that

$$
\begin{aligned}
&v_0^\lambda(x) = v_0^\infty(x) + \lambda^{-1} v_0^1(x) , \qquad \operatorname{div} v_0^\infty = 0 \\
&p_0^\lambda(x) = P_0 + \lambda^{-2} p_0^1(x) .
\end{aligned}
\tag{3.14}
$$

Thus, pressure variations should be much smaller than velocity variations. This completes the formal derivation of the asymptotic relationship between compressible and incompressible fluids. This is a singular limit because some of the characteristic speeds for the hyperbolic system in (3.3), the convective waves, stay bounded while other characteristic speeds, the sound waves, become infinite. The formal asymptotics just presented indicates the convergence to a system of nonlinear P.D.E.'s with infinite propagation speed of singular type, the incompressible Euler equations.

We remark here that the result of the formal asymptotic argument presented in (3.12) - (3.14) successfully predicts the leading order terms in the general expansion in (3.6); however, this formal derivation is not valid for the higher order terms. In fact, the correct asymptotic expansion to higher order is given by

$$
\begin{aligned}
p^\lambda &= P_0 + \lambda^{-2}(p^\infty(x,t) + p_1(x,t,\lambda)) + O(\lambda^{-3}) \\
v^\lambda &= v^\infty + \lambda^{-1}(v_1(x,t,\lambda)) + O(\lambda^{-2})
\end{aligned}
\tag{3.15}
$$

(see the discussion below and [16] for the rigorous proof), where $p_1(x,t,\lambda)$ is not zero generally and contains very fast scale acoustical oscillations - the same remarks apply to $v_1(x,t,\lambda)$. In fact, $p_1(x,t,\lambda)$, $v_1(x,t,\lambda)$ solve coupled equations of linear acoustics with variable coefficients defined by the incompressible background flow v^∞. A better formal expansion utilizing multiple time scales needs to be introduced at the outset to capture these higher order effects in the formal asymptotics.

With this formal derivation as motivation, we proceed to a rigorous treatment of the limiting process following the program outlined in (1.12). Our approach is completely elementary and uses only the classical tools of energy inequalities and Sobolev inequalities already developed in the previous sections. Several aspects of this singular limit were investigated earlier by Ebin [6], using a completely different approach via constrained infinite dimensional mechanical systems - we recommend this work for the interesting geometric insight that it provides. However, proofs to be discussed below yield much simpler proofs, stronger results for the limit of the equations in (3.5), and also unify the treatment of the compressible and incompressible Euler equations.

Uniform Stability for Compressible Fluid Flow Independent of Small Mach Numbers

First, we remark that a domain $\overline{G}_1 \subset\subset G$ of hyperbolicity in state space for the λ-dependent compressible fluid equations in (3.5) is defined by

$$\overline{G}_1 = \{(p^\lambda, v^\lambda) \,\big|\, |p^\lambda - P_0| \le \tfrac{P_0}{2},\ |v^\lambda| \le R'\}, \tag{3.16}$$

where $R' > 0$ is any convenient constant. We introduce

$$\begin{aligned} \tilde{p}^\lambda &= \lambda(p^\lambda - P_0) \\ u^\lambda &= {}^t(\tilde{p}^\lambda, v^\lambda) \end{aligned} \tag{3.17}$$

and compute from (3.5) that $\tilde{p}^\lambda$, v^λ satisfy the 4×4 quasi-linear symmetric hyperbolic system

$$\begin{aligned} \left(\gamma(P_0 + \lambda^{-1}\tilde{p}^\lambda)\right)^{-1} \frac{D\tilde{p}^\lambda}{Dt} + \lambda \operatorname{div} v^\lambda &= 0 \\ \rho(P_0 + \lambda^{-1}\tilde{p}^\lambda) \frac{Dv^\lambda}{Dt} + \lambda \nabla\tilde{p}^\lambda &= 0\,, \end{aligned} \tag{3.18 (a)}$$

with initial data

$$\begin{aligned} \tilde{p}^\lambda(x,0) &= \lambda(p_0^\lambda(x) - P_0) \\ v^\lambda(x,0) &= v_0^\lambda(x)\,. \end{aligned} \tag{3.18 (b)}$$

Since, as $\lambda \to \infty$, some of the coefficients of the hyperbolic system in (3.18) are becoming infinite, the time interval of local smooth existence $T(\lambda)$ determined in Theorem 2.1 might keep decreasing to zero as $\lambda \to \infty$. Here we prove that this catastrophy does not occur for a very wide class of initial data which easily includes the specific data in (3.14) - in fact, the pressure variations only need to be $O(\lambda^{-1})$ and the velocity variations can be $O(1)$. The proof relies on the specific balanced structure of the hyperbolic system in (3.18).

Theorem 3.1. (Uniform Stability of the Compressible Fluid Equations for Small Mach Numbers) Assume the inital data $(p_0^\lambda(x), v_0^\lambda(x))$ satisfy

$$\|\lambda(p_0^\lambda(x) - P_0)\|_{s_0} + \|v_0^\lambda(x)\|_{s_0} \le R \tag{3.19}$$

for some fixed $R > 0$, $\lambda \ge 1$, and $s_0 = [N/2] + 2$. Then there is a fixed time interval $[0,T_0]$, with $T_0 > 0$ and $\lambda_0(R)$, so that for $\lambda \ge \lambda_0(R)$ the compressible Euler equations in (3.5) have a classical solution on $[0,T_0]$ with

$$(\lambda(p^\lambda - P_0), v^\lambda) \in C([0,T_0], H^{s_0}) \cap C^1([0,T_0], H^{s_0-1})$$

and satisfying the estimate

$$|||\lambda(p^\lambda - P_0)|||_{s_0 T_0} + |||v^\lambda|||_{s_0,T_0} \le R' \tag{3.20}$$

with $R' > 0$ a fixed constant.

Remark. If in addition the initial data u^λ belongs to $H^s(R^N)$, then on the same interval $[0,T_0]$, u^λ belongs to $C([0,T_0], H^s) \cap C^1([0,T_0], H^{s-1})$ and satisfies

$$|||u^\lambda|||_{s,T_0} \le R'_s .$$

This follows directly from Corollary 1 of Theorem 2.2 and the proof of Lemma 2.4 below, but we have only stated Theorem 3.1 for the minimum number of derivatives s_0 for simplicity in expositon.

The crucial first observation necessary for the proof of the above Theorem 3.1 is the fact that with $u^\lambda = {}^t(\tilde{p}^\lambda, v^\lambda)$, the system in (3.18) can be rewritten as the symmetric hyperbolic system

$$\begin{aligned} A_0^\lambda \frac{\partial u^\lambda}{\partial t} + \sum_{j=1}^{N} A_j^\lambda \frac{\partial u^\lambda}{\partial x_j} &= 0 \\ u^\lambda(x,0) &= u_0^\lambda(x) \end{aligned} \tag{3.21}$$

where the coefficient matrices have the special structure

$$(3.22)\quad \begin{aligned} &(a) \quad A_j^\lambda = A_j(\lambda^{-1}u^\lambda, u^\lambda) + \lambda A_j^0, \qquad 1 \le j \le N \\ &(b) \quad A_0^\lambda = A_0(\lambda^{-1}u^\lambda) \end{aligned}$$

where

$$(3.23)\quad \begin{aligned} &(a) \quad A_j^0 \text{ are constant symmetric matrices} \\ &(b) \quad CI \le A_0(v) \le C^{-1}I, \qquad |v| \le \theta_0, \quad (c \le 1 \text{ w.l.o.g.}) \end{aligned}$$

and $A_j(v,u)$, $1 \le j \le N$, are smoothly varying for $|v| \le \theta_0$ and arbitrary values of $u \in R^M$. The structure in (3.22), (3.23) follows easily from (3.16), (3.18) where we have suppressed the constant P_0 in (3.22), (3.23). Furthermore, the conditions in (3.19) on the initial data become the requirement

$$\|u_0^\lambda(x)\|_{s_0} \le R .$$

The proof which we give for Theorem 3.1 contains an abstract stability theorem for the symmetric hyperbolic system in (3.21), satisfying the structural conditions in (3.22), (3.23) (see [16]). We will present the proof in this general context but leave the (obvious) statement which generalizes Theorem 3.1 to the interested reader. The proof of this general version of Theorem 3.1 proceeds via the continuation principle formulated in Theorem 2.2. The key lemma in this proof is the following one:

Lemma 3.1. Assume the structural conditions in (3.22) and (3.23) for the symmetric hyperbolic system in (3.21) and also

$$(3.24)\quad \|u_0^\lambda\|_{s_0} \le R .$$

Then there is a $\lambda_0(R)$ and constants $\tilde{M}_1(R,\theta_0), \tilde{M}_2(R,\theta_0)$ so that on any interval of classical existence $[0,T_*]$ for (3.21) where the additional a priori estimate

$$(3.25)\quad |||u^\lambda|||_{s_0,T_*} \le \frac{2R}{C}$$

is satisfied, we have for $\lambda \ge \lambda_0(R)$

$$(3.26)\quad \begin{aligned} &(a) \quad |\lambda^{-1}u^\lambda|_{L^\infty} \le \theta_0, \qquad 0 \le t \le T_* \\ &(b) \quad |Du^\lambda|_{L^\infty} \le \tilde{M}_1(R,\theta_0) \\ &(c) \quad \left|\frac{\partial}{\partial t} A_0^\lambda(u^\lambda)\right| \le \tilde{M}_2(R,\theta_0) . \end{aligned}$$

Furthermore, u^λ satisfies the energy estimate on $[0,T_*]$,

$$|||u^\lambda|||_{s_0,T_*} \le C^{-1} \exp(\tilde{C}(\tilde{M}_1 + \tilde{M}_2)T_*) \|u_0^\lambda\|_{s_0} \tag{3.27}$$

where $\tilde{C}$ depends only on θ_0 and λ_0.

WARNING: The conclusions in (3.26)(c) and (3.27) are not immediate corollaries of Theorem 2.2, although they will be proved in a similar fashion. The key fact regarding (3.27) is that $\tilde{C}$, $\tilde{M}_1$, $\tilde{M}_2$ are *independent of* λ. This does not follow directly from the estimate in (2.38) which might depend on λ through the constant C — we explicitly show that this potentially disastrous λ-dependence is not the case by re-examining the proof of Theorem 2.2 and applying the structural conditions in (3.22), (3.23).

We assume Lemma 3.1 for the moment and complete the proof of Theorem 3.1. With the fixed constants determined in (3.27), we define the fixed time interval $[0,T_0]$ independent of $\lambda \ge \lambda_0$ by the formula

$$2 = \exp(\tilde{C}(\tilde{M}_1 + \tilde{M}_2)T_0) . \tag{3.28}$$

Now, for any $\lambda \ge \lambda_0$ and any interval of local existence $[0,T_*]$ with $T_* \le T_0$, (3.24) and (3.27) together with (3.28) automatically imply the additional a priori estimate in (3.25). Since (3.25) implies (3.26)(a), (b), (c), which are the key hypotheses in the continuation principle formulated in Theorem 2.2, we deduce Theorem 3.1 by applying Theorem 2.2 for any fixed $\lambda \ge \lambda_0$. The estimate in (3.20) is an immediate consequence of the estimate in (3.27) and the choice of T_0 in (3.28).

It remains for us to prove Lemma 3.1 to complete the proof of Theorem 3.1. With the hypothesis in (3.25), we have for $0 \le t \le T_*$ by Sobolev's lemma

$$|\lambda^{-1}u^\lambda|_{L^\infty} \le \lambda^{-1} C_s |||u^\lambda|||_{s_0,T_*} \le \frac{2R}{\lambda C}$$

so that (3.26)(a) is guaranteed for $\lambda \ge \lambda_0$ with $\lambda_0 \equiv \frac{2R}{C\theta_0}$ - the proof of (3.26)(b) is an even simpler application of Sobolev's lemma. However, the proof of (3.26)(c) crucially utilizes the structural condition in (3.22), i.e., first we have

$$A_0^\lambda(u^\lambda) = A_0(\lambda^{-1}u^\lambda)$$

so that

$$\frac{\partial}{\partial t} A_0^\lambda(u^\lambda) = \left(\frac{\partial A_0}{\partial u}\right)\lambda^{-1}u_t^\lambda = -\frac{\partial A_0}{\partial u}\lambda^{-1}\sum_{j=1}^{N} A_j^\lambda \frac{\partial u^\lambda}{\partial x_j} . \tag{3.29}$$

Now, from (3.22)(a), $|A_j^\lambda| = O(\lambda)$, however these terms are multiplied by λ^{-1} in (3.29); thus, (3.26)(a), (b) and (3.29) imply the required estimate in (3.26)(c) with some constant $\tilde{M}_2(R_0)$. The final point in the proof of Lemma 3.1 is to establish the energy estimate in (3.27). Here, as in (2.41) - (2.45) from the proof of Theorem 2.2, we differentiate the equations and use the sharp calculus inequalities in Proposition 2.1, but here we crucially use the structural conditions in (3.22), (3.23). With the obvious notation for F_α^λ analogous to the notation in (2.41), (2.42), we compute that the dangerous commutator terms which might explode in λ have the form

$$\lambda A_0(\lambda^{-1}u^\lambda)[D^\alpha(A_0^{-1}(\lambda^{-1}u^\lambda)A_j^0 \frac{\partial u^\lambda}{\partial x_j}) - A_0^{-1}A_j^0 \frac{\partial}{\partial x_j} D^\alpha u^\lambda] . \tag{3.30}$$

We estimate the L^2 norm of the terms in (3.30) in the fashion already used in (2.43) by applying the calculus inequalities in (B), (C) of Proposition 2.1 - here, we make important use of (3.22), (3.23). We estimate

$$\begin{aligned} &\|\lambda A_0(\lambda^{-1}u^\lambda)[D^\alpha((A_0^\lambda)^{-1}A_j^0 u_{x_j}) - (A_0^\lambda)^{-1}A_j^0 D^\alpha u_{x_j}]\|_0 \\ &\le \tilde{C}\lambda[|D(A_0^{-1}(\lambda^{-1}u^\lambda)A_j^0)|_{L^\infty}\|D^{s_0-1}u_{x_j}\|_0 + \left|\frac{\partial u}{\partial x_j}\right|_{L^\infty}\|D^{s_0}(A_0^{-1}(\lambda^{-1}u^\lambda)A_j^0)\|_0] \\ &\le \tilde{C}|Du^\lambda|_{L^\infty}\|D^{s_0}u^\lambda\|_0 \end{aligned} \tag{3.31}$$

where the reader can check that we needed both the structural conditions in (3.22), (3.23) and the sharp calculus inequalities in Proposition 2.1 in the inequalities in (3.31). With the λ-independent estimates in (3.31) and the bounds in (3.26)(b), (c) which guarantee $|\operatorname{div} \vec{A}^\lambda|_{L^\infty} \le \tilde{C}(\tilde{M}_1 + \tilde{M}_2)$, the proof of the estimate in (3.27) of Lemma 3.1 proceeds in exactly the same fashion as the argument already given below (2.43) in the course of proving Theorem 2.2 - this completes the proof of Lemma 3.1.

Remark: The hypothesis $u_0^\lambda \in H^s$ *cannot be* generalized to $u_0^\lambda \in H_{u\ell}^s$ - the hyperbolic equations in (3.5) do not have a uniform-fixed finite speed of propagation.

Convergence of Compressible Fluids to Incompressible Fluids in the Zero Mach Number Limit – Constructive Existence for the Incompressible Euler Equations

Here, in a completely a priori fashion, we study the zero Mach number limit of the equations in (3.5) and give a constructive existence proof for classical solutions of the incompressible Euler equations in (3.2). Here we give the proof of the following

Theorem 3.2. Consider the solutions of the compressible fluid equations

$$(3.32)\qquad \begin{aligned} \gamma^{-1}(p^\lambda)^{-1}\frac{Dp^\lambda}{Dt} + \operatorname{div} v^\lambda &= 0 \\ \rho(p^\lambda)\frac{Dv^\lambda}{Dt} + \lambda^2\,\nabla p^\lambda &= 0 \end{aligned}$$

with the special initial data

$$(3.33)\qquad \begin{aligned} v^\lambda(x,0) &= v_0^\infty(x) + \lambda^{-1}v_0^1(x)\ , \qquad \operatorname{div} v_0^\infty = 0 \\ p^\lambda(x,0) &= P_0 + \lambda^{-2}p_0^1(x) \end{aligned}$$

belonging to H^{s_0} (and satisfying (3.19)). Let $[0,T_0]$ be the fixed time interval determined in Theorem 3.1. Then as $\lambda \to \infty$ there exists $v^\infty(x,t) \in L^\infty([0,T_0],H^{s_0})$ so that

$$(3.34)\qquad v^\lambda \to v^\infty(x,t) \quad \text{in} \quad C([0,T_0],H^{s_0-\varepsilon}_{loc})\ , \qquad \text{any } \varepsilon > 0\ .$$

The function $v^\infty(x,t)$ belongs to $C([0,T_0],H^{s_0}) \cap C^1([0,T_0],H^{s_0-1})$ and is a classical solution of the incompressible Euler equations, i.e., there exists p^∞ so that

$$(3.35)\qquad \begin{aligned} \rho_0\frac{Dv^\infty}{Dt} &= -\nabla p^\infty \\ \operatorname{div} v^\infty &= 0 \\ v^\infty(x,0) &= v_0^\infty(x) \end{aligned}$$

where $\rho_0 = \rho(P_0)$. The mean pressure p^∞ is the limit of p^λ in the weaker sense that

$$(3.36)\qquad \nabla p^\lambda \to \nabla p^\infty \quad \text{weak} * \text{ in} \quad L^\infty([0,T_0],H^{s-1})\ .$$

Before sketching the proof of the above theorem, we make the following comments:

Remark 1. The proof of Theorem 3.2 which we present uses the same number of derivatives as the one in [15], but one less derivative than the one in [16] - also, our proof of Theorem 3.2 is simpler than the one in [16] but uses the same ideas.

Remark 2. It follows from Theorem 3.1 that under the weaker hypothesis on the inital data in (3.19) there exists $\tilde{v} \in L^\infty([0,T_0],H^{s_0})$ so that

$$v^\lambda \to \tilde{v} \qquad \text{weak} * \text{ in} \quad L^\infty([0,T_0],H^{s_0})$$

by passing to a subsequence in λ. However, there is not sufficient compactness in time to conclude that $\tilde{v}$ solves the equation in (3.35) - the author suspects that v instead might solve some averaged equation which is not merely (3.35) in general. The reader can check that the time derivative of the fluid velocity $\frac{\partial v^\lambda}{\partial t}$ in (3.32) explodes as $\lambda \to \infty$ for the general initial data satisfying (2.19).

Given our Remark 2 above, we follow the general strategy first expounded by Lions [20] and seek an estimate which guarantees compactness in time in some very low norm. As we shall see below, this requirement leads naturally to the "slightly compressible" conditions in (3.33) for the initial data. As in Theorem 3.1, there is an abstract version of Theorem 3.2, so we revert to the notation from (3.21),

$$A_0^\lambda \frac{\partial u^\lambda}{\partial t} + \sum_{j=1}^{N} A_j^\lambda \frac{\partial u^\lambda}{\partial x_j} = 0 \tag{3.37}$$
$$u^\lambda(x,0) = u_0^\lambda(x) ,$$

where the structural conditions in (3.22), (3.23) are satisfied and make the identification $u^\lambda = {}^t(\lambda(p^\lambda - P_0), v^\lambda)$. By differentiating (3.37) with respect to time, we compute that $v^\lambda \equiv u_t^\lambda$ satisfies the hyperbolic system

$$A_0^\lambda \frac{\partial v^\lambda}{\partial t} + \sum_{j=1}^{N} A_j^\lambda \frac{\partial v^\lambda}{\partial x_j} + B^\lambda v^\lambda = 0 \tag{3.38}$$
$$v^\lambda(x,0) = - A_0^{-1}(\lambda^{-1}u_0^\lambda)\Big(\sum_{j=1}^{N} A_j^\lambda(u_0^\lambda) \frac{\partial u_0^\lambda}{\partial x_j}\Big)$$

where B^λ is given by

$$B^\lambda \equiv \lambda^{-1} \frac{\partial A_0}{\partial u}\Big|_{\lambda^{-1}u^\lambda} + \sum_{j=1}^{N} \Big[\frac{\partial}{\partial u} A_j^\lambda(\lambda^{-1}u^\lambda, u^\lambda)\Big] u_{x_j} . \tag{3.39}$$

Once again, it follows from the specific structural conditions in (3.22), (3.23) and Theorem 3.1 that

$$|B^\lambda(t)|_{L^\infty} \le \tilde{C} , \qquad 0 \le t \le T_0 . \tag{3.40}$$

Now, with the information in (3.40) and (3.22), (3.23), we apply the energy principle from (1.21) to v^λ to guarantee the estimate

$$|||u_t^\lambda|||_{0,T_0} \le C\|u_t^\lambda(0)\|_0 \tag{3.41}$$

for $\lambda \ge \lambda_0(R)$. Thus we have the estimate

$$|||u_t^\lambda|||_{0,T_0} \le R' , \tag{3.42}$$

provided that the norm $\|u_t^\lambda(0)\|_0$ stays bounded independent of λ; this will be the case provided that the condition

$$\|\lambda \sum_{j=1}^{N} A_j^0 \frac{\partial}{\partial x_j} (u_0^\lambda(x))\|_0 \leq \tilde{R} \tag{3.43}$$

is satisfied for some constant $\tilde{R} > 0$. For the concrete system of equations in (3.18) associated with compressible fluid flow, the "slightly compressible" initial conditions in (3.33) guarantee the bound in (3.43).

Returning to the concrete equations in (3.32), we deduce the estimates

$$\lambda|||p_t^\lambda|||_{0,T_0} + |||v_t^\lambda|||_{0,T_0} \leq R' \tag{3.44}$$

from the above argument provided the initial data has the form in (3.33). From Theorem 3.1 we also have the general estimate

$$\lambda|||p^\lambda - P_0|||_{s_0,T_0} + |||v^\lambda|||_{s_0,T_0} \leq R' . \tag{3.45}$$

Now, by applying the Arzela-Ascoli theorem, the interpolation inequalities in (2.20), and passing to a suitable subsequence in a familiar fashion, we conclude from (3.44) and (3.45) that there exists $v^\infty \in L^\infty([0,T_0],H^{s_0})$ so that

$$v^\lambda \to v^\infty \quad \text{in} \quad C([0,T],H_{loc}^{s_0-\varepsilon}) \tag{3.46}$$

for any $\varepsilon > 0$. Furthermore, from (3.44), (3.45), we also conclude that $v^\infty \in \mathrm{Lip}([0,T_0], L^2) \cap C([0,T_0],C^1(R^N))$.

In our next step we identify the limit v^∞ as a solution of the incompressible Euler equations. By using the estimates in (3.44), (3.45) in the pressure equation from (3.32), we see that

$$C(R') \geq \lambda|||\gamma^{-1}(p^\lambda)^{-1} \frac{Dp^\lambda}{Dt}|||_{0,T_0} = \lambda|||\mathrm{div}\ v^\lambda|||_{0,T_0} .$$

Therefore, (3.46) and the above estimate imply

$$\mathrm{div}\ v^\infty = 0 . \tag{3.47}$$

Next, we deduce that v^∞ satisfies a weak form of the first equations in (3.35). Let w be an arbitrary vector-valued function with $w \in H^{s_0}$, w rapidly decreasing, and $\mathrm{div}\ w = 0$; let $\phi(t) \in C_0^\infty((0,T_0))$ be any smooth scalar-valued test function. We recall that $(w,v)_0$ denotes the L^2 inner product and also from (3.7) that

$$(w, \lambda^2 \nabla p^\lambda)_0 = 0 .$$

Taking the inner product of the second equation in (3.32) with w and applying the test function $\phi(t)$, we obtain

$$(3.48)\qquad 0 = \int_0^{T_0} -\frac{\partial\phi}{\partial t}(w,\rho^\lambda v^\lambda)_0 + \phi(w,\rho^\lambda v^\lambda\cdot\nabla v^\lambda)_0 dt - \lambda^{-1}\int_0^{T_0}\phi(t)(w,\frac{\partial\rho}{\partial p}(p^\lambda)\frac{\partial\tilde{p}^\lambda}{\partial t}v^\lambda)_0\,dt$$

where $\rho^\lambda = \rho(p^\lambda)$. From (3.44) and (3.45), we derive as $\lambda\to\infty$

$$\rho^\lambda \to \rho(P_0) = \rho_0 \quad \text{uniformly on} \quad [0,T_0]\times R^N$$

$$|||\frac{\partial\tilde{p}^\lambda}{\partial t}v^\lambda|||_{0,T_0} \le C(R')$$

and also from (3.46) and the dominated convergence theorem (recall w is rapidly decreasing),

$$(w,\rho^\lambda v^\lambda\cdot\nabla v^\lambda)_0 \to (w,\rho_0 v^\infty\cdot\nabla v^\infty)_0$$

$$(w,\rho^\lambda v^\lambda)_0 \to (w,\rho_0 v^\infty)_0$$

uniformly on $[0,T_0]\times R^N$ so that passing to the limit in (3.48), we have

$$(3.49)\qquad 0 = \int_0^{T_0}[-\frac{\partial\phi}{\partial t}(w,\rho_0 v^\infty)_0 + \phi(w,v^\infty\cdot\nabla v^\infty)_0]dt$$

for all $\phi\in C_0^\infty((0,T_0))$ and $w\in H^s$ with $\operatorname{div} w = 0$ and w rapidly decreasing. (We leave it as an exercise to the reader to check that rapidly decreasing smooth w with $\operatorname{div} w = 0$ are dense in the space of all $w\in L^2$ satisfying this condition.) Thus, (3.47) and (3.49) identify v^∞ as a weak solution of the incompressible Euler equations in (3.35) with $v^\infty(x,0) = v_0^\infty(x)$ and $v^\infty\in \mathrm{Lip}([0,T_0],L^2\cap C([0,T_0],C^1)\cap L^\infty([0,T_0],H^s)$. Now, $v^\infty\in L^\infty([0,T_0],H^{s_0})$ and $s_0 = [n/2]+2$ imply that

$$\frac{1}{\rho_0}P(v^\infty\cdot\nabla v^\infty)\in L^\infty([0,T_0],H^{s_0-1})$$

and since the weak form in (3.49) implies that

$$\frac{\partial v^\infty}{\partial t} = \frac{1}{\rho_0}P(v^\infty\cdot\nabla v^\infty)$$

in the sense of distributions, we conclude that $v^\infty\in \mathrm{Lip}([0,T_0],H^{s_0-1})\cap L^\infty([0,T_0],H^{s_0})$. Since v^∞ already belongs to $C([0,T_0],C^1(R^N))\cap \mathrm{Lip}([0,T_0],H^{s_0-1})\cap L^\infty([0,T_0],H^{s_0})\cap C([0,T_0],H^{s'}_{loc})$, $s'<s_0$, it is not difficult to prove that

$$\frac{1}{\rho_0}P(v^\infty\cdot\nabla v^\infty)\in C([0,T_0],C_{loc}(R^N))$$

and therefore to deduce that $\frac{\partial v^\infty}{\partial t} \in C([0,T_0] \times R^N)$ so that $v^\infty \in C^1([0,T_0] \times R^N)$ and is a classical solution of (3.35). For brevity, we leave the details to the reader. The convergence of the pressures is discussed in [15] so we won't repeat it here. The additional regularity $v^\infty \in C([0,T],H^{s_0}) \cap C^1([0,T],H^{s_0-1})$ follows from the weaker regularity already established and an argument like the one in Theorem 2.1(b) - once again, we omit the details. A number of details in the convergence are simpler in the periodic case (see [15]) - we have to be more careful here because of the behavior at infinity.

A Continuation Principle for the Incompressible Euler Equations

In the previous subsection we constructed classical solutions

$$v^\infty \in C([0,T],H^s) \cap C^1([0,T],H^{s-1})$$

for the incompressible Euler equations

$$\frac{Dv^\infty}{Dt} + \frac{1}{\rho_0}\nabla p^\infty = 0$$

$$\text{div } v^\infty = 0 \tag{3.50}$$

$$v^\infty(x,0) = v_0^\infty(x) \ ,$$

$s > \frac{N}{2} + 1$, where $v_0^\infty \in H^s(R^N)$. Here we characterize the maximal interval $[0,T_*)$ of classical H^s local existence for the solutions of (3.50). We have the following result:

Theorem 3.3. The interval $[0,T_*)$ with $T_* < \infty$ is a maximal interval of H^s local existence if and only if

$$\max_{1\le i,j\le N} \left|\frac{\partial v_i^\infty}{\partial x_j}\right|_{L^\infty} \to +\infty \quad \text{as} \quad t \uparrow T_* . \tag{3.51}$$

In particular, we have global existence of an H^s solution to the Euler equations provided that the first derivatives of the velocity satisfy the a priori estimate

$$\max_{1\le i,j\le N} \left|\frac{\partial v_i}{\partial x_j}\right|_{L^\infty}(t) \le M(t)$$

with $M(t) \in C([0,\infty))$.

The proof uses energy identities as in Theorem 2.2 but is much simpler because the nonlinearity is quadratic and the equations in (3.50) are defined for all $v^\infty \in R^N$ (i.e., $G = R^N$). With $v_\alpha^\infty = D^\alpha v^\infty$ and $|\alpha| \le s$, we compute that

(3.52) $$\frac{\partial v_\alpha^\infty}{\partial t} + v^\infty \cdot \nabla v_\alpha^\infty + \frac{1}{\rho_0} \nabla p_\alpha^\infty = F_\alpha$$

with

$$F_\alpha^i = - \sum_{j=1}^{N} [D^\alpha (v_j \frac{\partial v_i}{\partial x_j}) - v_j \, D^\alpha (\frac{\partial v_i}{\partial x_j})] \; .$$

Now, applying the calculus inequality from Proposition 2.1 (B), we obtain

$$\sum_{1\le|\alpha|\le s} \|F_\alpha\|_0 \le C_s \max_{1\le i,j\le N} \left|\frac{\partial v_i^\infty}{\partial x_j}\right|_{L^\infty} \|D^s v^\infty\|_0 \; .$$

Next, multiplying the equation in (3.52) by v_α^∞ and integrating over R^N, we obtain

$$\sum_{0\le|\alpha|\le s} \frac{\partial}{\partial t} (v_\alpha^\infty, v_\alpha^\infty)_0 = 2 \sum_{1\le|\alpha|\le s} (F_\alpha, v_\alpha^\infty) \le 2C_s \max_{1\le i,j\le N} \left|\frac{\partial v_i}{\partial x_j}\right|_{L^\infty} \|v^\infty\|_s \; ,$$

so that by Gronwall's inequality we have the estimate

(3.53) $$|||v^\infty|||_{s,T} \le \exp\Big(2C_s \max_{1\le i,j\le N} \left|\frac{\partial v_i^\infty}{\partial x_j}\right|_{L^\infty} T\Big) \|v_0^\infty\|_s$$

on any interval $[0,T]$ of classical H^s local existence. With the estimate in (3.53), we deduce Theorem 3.3 by applying the same argument as we used in Theorem 2.2.

Theorem 3.3 is a superficial result regarding a characterization of the breakdown time for the Euler equations; however, this result still improves those existing in the published literature. The physical case $N = 3$ is especially interesting. Here the 3×3 matrix $(\frac{\partial v_i}{\partial x_j})$ has its antisymmetric part

$$\frac{1}{2} \left(\frac{\partial v_i}{\partial x_j} - \frac{\partial v_j}{\partial x_i}\right) \equiv \Omega \; ,$$

where Ω *is* essentially *the vorticity* and symmetric part given by

$$\frac{1}{2} \left(\frac{\partial v_i}{\partial x_j} + \frac{\partial v_j}{\partial x_i}\right) = D \; ,$$

where D is the deformation tensor. Given the three independent components of the vorticity matrix Ω, and the five independent components of the deformation tensor D, which of these components control breakdown? One might conjecture on the basis of numerical evidence that if $|\Omega|_{L^\infty}$ stays bounded, then automatically global existence of H^s solutions is obtained, i.e., the behavior of the vorticity matrix alone characterizes the maximal interval of H^s classical existence. This is the situation for $N = 2$ (see recent work by daVeiga [27]). An outstanding open problem is to find H^s initial data for the incompressible Euler equations in $N = 3$ where finite time breakdown occurs. The extremely interesting recent numerical evidence of Chorin

[2] indicates that the set where such a breakdown occurs may be enormously complex. However, Theorem 3.3 indicates that, as in the phenomenon of shock formation, the maximum norm of the first derivatives of velocity is the important quantity controlling this breakdown. Similar results are valid in the periodic case.

The Asymptotic Expansion of Compressible Fluid Flow at Low Mach Numbers Via Incompressible Flow and Linearized Acoustics

Here, we state the rigorous results on the asymptotic expansion of solutions of the compressible fluid equations in (3.32) with the initial data

$$v^\lambda(x,0) = v_0^\infty(x) + \lambda^{-1} v_0^1(x), \qquad \operatorname{div} v_0^\infty = 0$$

$$p^\lambda(x,0) = P_0 + \lambda^{-2} p_0^1(x)$$

which are proved in the second half of [16]. Given the incompressible limit solution (v^∞, p^∞) which satisfies the incompressible Euler equations in (3.35), we let $(v_1^\lambda, p_1^\lambda)$ be the solution of the *linearized equations of acoustics*,

$$(3.54) \qquad \begin{aligned} &(\gamma P_0)^{-1} \frac{D}{Dt} p_1^\lambda + \lambda \operatorname{div} v_1^\lambda = F_1 \\ &\rho_0\left(\frac{D}{Dt} v_1^\lambda + (v_1^\lambda \cdot \nabla) v^\infty\right) + \lambda \nabla p_1^\lambda = 0 \end{aligned}$$

with the initial data

$$(3.55) \qquad (p_1^\lambda(x,0), v_1^\lambda(x,0)) = (-p^\infty(x,0) + p_0^1(x), v_0^1(x))$$

and F_1 defined by

$$(3.56) \qquad F_1 = -(\gamma P_0)^{-1} \frac{Dp^\infty}{Dt},$$

where in (3.54) $\frac{D}{Dt} = \frac{\partial}{\partial t} + \sum_{j=1}^{N} v_j^\infty \frac{\partial}{\partial x_j}$.

When $v^\infty \equiv 0$, the equations in (3.54) reduce to the familiar equation of linear acoustics

$$(3.57) \qquad \begin{aligned} &(\gamma P_0)^{-1} \frac{\partial p_1^\lambda}{\partial t} + \lambda \operatorname{div} v_1^\lambda = 0 \\ &\rho_0 \frac{\partial v_1^\lambda}{\partial t} + \lambda \nabla p_1^\lambda = 0 \\ &p_1^\lambda(x,0) = p_0^1(x), \qquad v_1^\lambda(x,0) = v_0^1(x). \end{aligned}$$

In this special case, we use (3.7) to decompose v_1^λ, $v_0^1(x)$ as

$$v_1^\lambda = w_1^\lambda + \nabla\psi_1^\lambda , \qquad v_0^1(x) = w_0^1(x) + \nabla\psi_0^1(x)$$

where $\operatorname{div} w_1^\lambda = 0$, $\operatorname{div} w_0^1 = 0$. With the above decomposition, we compute that solutions of (3.57) satisfy the equivalent equations

$$(3.58)\quad \begin{aligned} &(a) && w_1^\lambda(x,t) \equiv w_0^1(x) \\ &(b) && \frac{1}{\lambda}\rho_0 \frac{\partial\psi_1^\lambda}{\partial t} = -p_1^\lambda \\ &(c) && \frac{\partial^2\psi_1^\lambda}{\partial t^2} - c_0^2\lambda^2\nabla\psi_1^\lambda = 0 , \qquad c_0^2 = \frac{\gamma P_0}{\rho_0} \\ &(d) && \psi_1^\lambda(x,0) = \psi_0^1(x) , \qquad \frac{1}{\lambda}\rho_0 \frac{\partial\psi_1^\lambda}{\partial t}\bigg|_{t=0} = -p_0^1(x) . \end{aligned}$$

The equation in (3.58)(c) is the famous wave equation of linear acoustics. In general, the equations in (3.54) are linear variable coefficient acoustic equations containing both slow and fast speeds of propagation, as illustrated in the special case described in (3.57), (3.58).

The intuitive content of the theorem to be stated below is the following fact: For small Mach numbers and $0 \le t \le T_0$,

$$(3.59)\quad \begin{aligned} p^\lambda(x,t) &= P_0 + \lambda^{-2}(p_1^\lambda(x,t) + p^\infty(x,t)) + O(\lambda^{-3}) \\ v^\lambda(x,t) &= v^\infty(x,t) + \lambda^{-1}(v_1^\lambda(x,t)) + O(\lambda^{-2}) \end{aligned}$$

where (v^∞, p^∞) satisfy the incompressible fluid equations and $(v_1^\lambda, p_1^\lambda)$ solve the linear equations of acoustics in (3.54) with initial data in (3.55) and forcing function F_1 defined in (3.56).

Theorem 3.4. Assume that a solution of the incompressible fluid equations in (3.35) exists on some interval $[0,T_0]$, and satisfies for some $s > \frac{n}{2} + 1$,

$$(3.60)\quad |||v^\infty|||_{s+1,T_0} + |||v_t^\infty|||_{s,T_0} + |||p^\infty|||_{s+2,T_0} + |||p_t^\infty|||_{s+1,T_0} \le C(T_0) .$$

Then for $\lambda \ge \lambda_0$, the compressible fluid equation in (3.32) with the initial data in (3.33) have a solution (p^λ, v^λ) in $C([0,T_0],H^s) \cap C^1([0,T_0],H^{s-1})$. Furthermore, there is a constant $C > 0$ so that

$$(3.61)\quad \begin{aligned} |||v^\lambda - (v^\infty + \lambda^{-1}v_1^\lambda)|||_{s,T_0} &\le C\lambda^{-2} \\ |||p^\lambda - \lambda^{-2}(p^\infty + p_1^\lambda)|||_{s,T_0} &\le C\lambda^{-3} , \end{aligned}$$

i.e., the asymptotic expansion described in (3.59) is convergent in the norms described in (3.61).

Remark. For functions defined on R^N, the requirement

$$|||p^{\infty}|||_{0,T_0} + |||p_t^{\infty}|||_{0,T_0} \leq C(T_0)$$

is a genuine restriction on the velocity field beyond the requirement that $v \in H^s$ - this requires the velocity to vanish sufficiently rapidly at infinity. On the other hand, Theorem 3.4 also applies in the periodic case where (3.60) provides no new restrictions.

Some open Problems Regarding Singular Limits in Fluid Dynamics and Related Topics

Here we mention several open problems related to the singular limits mentioned in this subsection. There are open questions in three main areas which we describe below (including any partial progress on these questions known to the author).

TOPIC 1. *Singular Limits in Whole Space or Periodic Problems*

(A) The first problem we describe is one for the equations of linear acoustics in (3.54). This is a problem for *linear hyperbolic equations* with variable coefficients; namely, determine the *detailed* asymptotic representation for solutions of (3.54) as $\lambda \to \infty$. For the whole space, scattering of waves may occur and/or ray trapping; for the periodic case, waves may be *averaged* in specific ways related to the coefficients or resonance might occur too.

(B) For sufficiently small initial data, Nishida and Matsumura [23] have proved the global existence of classical solutions for the full-compressible Navier-Stokes equations at fixed Mach number. Investigate the global convergence to classical solutions of the incompressible Navier-Stokes equations as the Mach number tends to zero. A. Benabdallah has described some very nice results at this meeting, which settle the above problem for the "artificial" compressibility method of Chorin and Temam. Perhaps her work generalizes to the full Navier-Stokes system.

Remark. When entropy variations or equivalently temperature variations are extremely small and heat conduction vanishes, the results in [15] due to Klainerman and the author prove that for large initial fluid velocities the low Mach number limit of the compressible Navier-Stokes equations is the incompressible Navier-Stokes equations. However, when *temperature variations* are *large* and heat conduction coefficients are nonzero, the *singular limit system for the compressible Navier-Stokes equations* as the Mach number vanishes *is* a system resembling the nonhomogeneous fluid equations coupled nonlinearly to the convective heat conduction equation and *not the incompressible Navier-Stokes equations.* S. Klainerman and the author are developing these results currently - the proofs in this case are more subtle and require some new estimates beyond those discussed in these notes or [15], [16].

TOPIC 2. *Singular Limits in Domains with Boundaries*

(A) D. Ebin [7] has recently given a proof of the singular limit of the isentropic compressible fluid equations in bounded domains using the equations in Lagrangian coordinates - this proof is too complex in the author's opinion. Find a direct classical proof of the convergence following the strategy discussed in these lectures for the whole space involving Friedrichs' theory of symmetric systems. The condition requiring the normal velocity to vanish at the boundary creates a technical difficulty in adopting the proofs in a straightforward fashion. However, recently Rauch and Nishida (private communication) have succeeded in adapting the conventional approach to the equations at fixed Mach number in a bounded domain. Perhaps their methods can be modified to handle the singular limit as well.

(B) We ask the same question as in (A) for the full compressible Navier-Stokes equations (with small temperature variations to start with) and *fixed* viscosity (to avoid difficulties with boundary layers). Here the phenomena should be different than in (A) since the type of the equations has changed and the no-slip boundary conditions are better behaved for the parabolic operators in these equations. Temam's work [26] settles this question for *weak solutions* of the equations of "artificial" compressibility - this is an excellent starting point for this problem. Perhaps one would begin by studying the convergence in bounded domains in the *classical* solution regime for artificial compressibility - the early work of Chorin [28] should be helpful here. Temam's results for the artificial compressibility method also suggest the following problem: Study the singular limits for *weak* solutions of the viscous fluid equations. For the whole space or periodic problems, this question is still quite interesting.

(C) Convergence of compressible viscous solutions to incompressible viscous solutions for the *steady* equations in the low Mach number regime - once again both in all of space and bounded domains. The simplest problem of this type is steady potential flow in 2-D in the whole space for the inviscid equations - one only needs to prove a nonlinear perturbation theorem for solutions of the Laplace equation under suitable conditions on the data. At this meeting, M. Padula has informed us of some partial results for the steady viscous case at low Reynolds number with other additional hypotheses.

(D) Develop the detailed higher order asymptotics for the viscous case in a bounded domain. With this information, we ask the same question as in Topic 1 (A) for the associated linear system. There are some first results due to Geymonat and Sanchez-Palencia [9] on the linearized problem.

TOPIC 3. *Nonlinear Singular Limits in other Physical Problems*

Besides the discussion of singular limits to be given in the next section, there are two other very interesting physical systems of equations with a myriad of singular limits of the type discussed here (and probably many more physical problems unknown to this author!!)

(A) The equations of meteorology (see [30]).

(B) The equations of magneto-fluid dynamics (some first results are described in [15]).

Remark: In interesting work, Kreiss and his co-workers [18] study such singular limits for *special* initial data so that *many* higher-time derivatives *stay bounded* and the general phenomena studied here are completely suppressed. The problems for general initial data are largely wide open for the systems in (A), (B) above - especially, as regards a priori proofs.

4. Equations for Low Mach Number Combustion

Here we develop the appropriate singular limit equations for low Mach number combustion. For simplicity in exposition, we consider this system of equations at infinite Reynolds' number so that $\tilde{v} = 0$ - the derivation we present below also applies when these terms are included (see [21]). With this assumption, these equations can be written as we have done previously in the nondimensional form

$$(4.1)\quad \begin{aligned} &(a)\quad \frac{Dp^\lambda}{Dt} + \gamma p^\lambda \operatorname{div} v^\lambda = [\operatorname{div}(k\nabla T^\lambda) + q_0 \rho^\lambda W(z^\lambda, T^\lambda)] \\ &(b)\quad \rho^\lambda \frac{Dv^\lambda}{Dt} + \lambda^2 \nabla p^\lambda = 0 \\ &(c)\quad \rho^\lambda \frac{DT^\lambda}{Dt} + (\gamma-1) p^\lambda \operatorname{div} v^\lambda = [\operatorname{div}(k\nabla T^\lambda) + q_0 \rho^\lambda W(z^\lambda, T^\lambda)] \\ &(d)\quad \rho^\lambda \frac{Dz^\lambda}{Dt} = \operatorname{div}(\rho^\lambda d\, \nabla z^\lambda) - \rho^\lambda W(z^\lambda, T^\lambda) \end{aligned}$$

with p^λ the pressure, T^λ the temperature, z^λ the mass fraction of unburnt gas

$$(4.2)\qquad \rho^\lambda = \frac{p^\lambda}{T^\lambda}, \qquad W(z,T) \equiv z\, K_0\, e^{-A_0/T}$$

and $\lambda^2 = M^{-2}/\gamma A$ with M the Mach number. (We have set the heat capacity c_p to be one.)

In the formal derivation to be presented below, we assume that the physical quantities in (4.2) are defined in a region Ω, where either

$$(4.3)\qquad \begin{aligned} &(a)\quad \overline{\Omega} \text{ is a bounded domain} \\ &\qquad \text{or} \\ &(b)\quad \overline{\Omega} \text{ is the whole space.} \end{aligned}$$

In the case of (4.3)(a), we have combustion in a confined chamber and the qualitative effects are quite different from those in the whole space, as we shall see below. For bounded domains we assume the boundary conditions on the boundary $\partial\Omega$ are given by

$$(4.4)\qquad \begin{aligned} &(a)\quad v^{\lambda}\cdot\vec{n}\big|_{\partial\Omega} = 0 \\ &(b)\quad \frac{\partial T^{\lambda}}{\partial n}\Big|_{\partial\Omega} = 0\,, \qquad \frac{\partial z^{\lambda}}{\partial n}\Big|_{\partial\Omega} = 0\,. \end{aligned}$$

The first boundary condition in (4.4) is the familiar solid wall boundary condition from fluid dynamics, while those in (4.4)(b) for T^{λ}, z^{λ} guarantee the boundary is insulated and also that no new unburnt gas is injected at the boundary - these conditions from (4.4)(b) are imposed only for simplicity in exposition. In the case of bounded domains, we use the following analogue of (3.7),

$$(4.5)\qquad \begin{aligned} &\text{Every } v \in L^2(\overline{\Omega}) \text{ admits the unique orthogonal decomposition} \\ &\qquad\qquad v = w + \nabla\phi \\ &\text{where } \operatorname{div} w = 0,\quad w\cdot\vec{n}\big|_{\partial\Omega} = 0\,. \end{aligned}$$

With these preliminary facts, we give the formal derivation of the limit equations for (4.1) and concentrate on the case of a bounded region. We begin with the Ansatz

$$(4.6)\qquad \begin{aligned} p^{\lambda} &= p^{\infty} + \lambda^{-1}p_1 + \lambda^{-2}p_2 + O(\lambda^{-3}) \\ v^{\lambda} &= v^{\infty} + \lambda^{-1}v_1 + O(\lambda^{-2}) \\ T^{\lambda} &= T^{\infty} + \lambda^{-1}T_1 + O(\lambda^{-2}) \\ z^{\lambda} &= z^{\infty} + \lambda^{-1}z_1 + O(\lambda^{-2})\,. \end{aligned}$$

By substituting the expansion from (4.6) into the velocity equations and equating powers of order two, one, and zero (after applying the projection), we obtain as in (3.8), (3.9),

(4.7)
(a) $P^\infty \equiv P^\infty(t)$

(b) There is a scalar pressure p^∞ so that $\rho(P^\infty,T^\infty)\,\frac{Dv^\infty}{Dt} = -\nabla p^\infty$

and $\rho(P^\infty,T^\infty) = P^\infty/T^\infty$ (generally not constant!). The equations of order zero in (4.1)(c), (d) for T^∞, z^∞ are completely straightforward and given by

(4.8)
$$\text{(a)}\quad \rho^\infty \frac{DT^\infty}{Dt} + (\gamma-1)P^\infty \operatorname{div} v^\infty = \operatorname{div}(k\nabla T^\infty) + q_0\rho^\infty W(z^\infty,T^\infty)$$
$$\text{(b)}\quad \rho^\infty \frac{Dz^\infty}{Dt} = \operatorname{div}(\rho^\infty d\nabla z^\infty) - \rho^\infty W(z^\infty,T^\infty)$$

with $\rho^\infty = P^\infty/T^\infty$. The only subtle point is the equation for P^∞ which we now discuss; the order zero terms in the pressure equation are given by

$$(4.9)\qquad (\gamma P^\infty)^{-1}\frac{dP^\infty}{dt} = -\operatorname{div} v^\infty + (P^\infty\gamma)^{-1}[\operatorname{div}(k\nabla T^\infty) + q_0\rho^\infty W(z^\infty,T^\infty)]\,.$$

Now, the left-hand side of (4.9) is a function of time alone, while the right-hand side involves functions of both space and time. Thus, the self-consistency of (4.9) to order zero demands that we find a scalar function $H(t)$, so that

(4.10)
$$\text{(a)}\quad (\gamma P^\infty)^{-1}\frac{dP^\infty}{dt} = H(t)$$
$$\text{(b)}\quad \operatorname{div} v^\infty = -H(t) + (P^\infty\gamma)^{-1}[\operatorname{div}(k\nabla T^\infty) + q_0\rho^\infty W(z^\infty,T^\infty)]$$

The key observation is that both equation (a) and equation (b) can be satisfied simultaneously by a unique choice of $H(t)$. To see this, we use the decomposition in (4.5) and write $v^\infty = w^\infty + \nabla\psi^\infty$. The solid wall boundary condition for v^∞ from (4.4) and (4.5), (4.10)(b) give

(4.11)
$$\Delta\psi^\infty = G(x,t) \qquad \text{in } \overline{\Omega}$$
$$\left.\frac{\partial\psi^\infty}{\partial\vec{n}}\right|_{\partial\Omega} = 0$$

where $G(x,t)$ is defined by the right-hand side of (4.10)(b). Now, the elliptic equation in (4.11) has a solution with $\nabla\psi^\infty$ uniquely determined if and only if

$$(4.12)\qquad \int_\Omega G(x,t)\,dx = 0\,.$$

The equation in (4.12) together with the boundary condition in (4.4)(b) uniquely determines $H(t)$ by the nonlocal quantity

$$H(t) \equiv \frac{q_0\gamma^{-1}}{\mathrm{Vol}(\Omega)} \int_\Omega (T^\infty)^{-1} W(z^\infty,T^\infty)dx . \tag{4.13}$$

With $v^\infty = w^\infty + \nabla\psi^\infty$, we summarize the derivation in (4.6) through (4.13) with the following system of <u>*Equations for Low Mach Number Combustion:*</u>

(4.14)

(a) $$(P^\infty)^{-1} \frac{dP^\infty}{dt} = \frac{q_0}{\mathrm{Vol}(\Omega)} \int_\Omega (T^\infty)^{-1} W(z^\infty,T^\infty)dx$$

(b) $$\Delta\psi^\infty = - \frac{q_0}{\mathrm{Vol}(\Omega)} \int_\Omega (T^\infty)^{-1} W(z^\infty,T^\infty)dx + (P^\infty\gamma)^{-1}[\mathrm{div}(k\nabla T^\infty) + q_0\rho^\infty W(z^\infty,T^\infty)] ,$$

$$\left.\frac{\partial\psi^\infty}{\partial n}\right|_{\partial\Omega} = 0$$

(c) $$\frac{P^\infty}{T^\infty}\left(\frac{D}{Dt} w^\infty\right) = - \frac{P^\infty}{T^\infty}\left(\frac{D}{Dt}\nabla\psi^\infty\right) - \nabla p^\infty , \qquad \mathrm{div}\, w^\infty = 0 , \qquad w^\infty\cdot n\big|_{\partial\Omega} = 0$$

(d) $$\rho^\infty\left(\frac{Dz^\infty}{Dt}\right) = \mathrm{div}(\rho^\infty d\nabla z^\infty) - \rho^\infty W(z^\infty,T^\infty) , \qquad \left.\frac{\partial z^\infty}{\partial n}\right|_{\partial\Omega} = 0$$

(e) $$\rho^\infty \frac{DT^\infty}{Dt} = \frac{\gamma-1}{\gamma}\frac{dP^\infty}{dt} + \frac{1}{\gamma}\mathrm{div}(\kappa\nabla T^\infty) + \frac{1}{\gamma} q_0\rho^\infty W(z^\infty,T^\infty), \qquad \left.\frac{\partial T^\infty}{\partial n}\right|_{\partial\Omega} = 0 .$$

While the system in (4.14) might seem rather complex at first glance, it is quite susceptible to numerical computation via the method of fractional steps because in successive steps,

(4.15)

(a) is a scalar nonlinear O.D.E.

(b) is an elliptic equation

(c) is a nonhomogeneous incompressible flow equation

(d), (e) is a system of parabolic reaction-diffusion equations.

In fact, a numerical algorithm in unbounded regions which uses the vortex method in step (c) and solves (b) completely while utilizing the Landau approximation of flame theory in steps (d), (c) has been developed in [29] for such a purpose. We recommend [29] both for a description of the algorithm and computational results. For bounded domains, one only needs to incorporate (a) as a separate fractional step - this is a trivial matter. The equations in (4.14) define a well-posed problem; in a Ph.D. dissertation currently being written under the author's supervision, P. Embid is developing these as well as other rigorous mathematical properties of the equations in (4.14). For a bounded domain and an exothermic reaction so that $q_0 > 0$, we see that the equation in (4.14)(a) predicts *a rise in pressure* due to heat release - this is physically reasonable. On the other hand, when Ω is the whole space, the equation in (4.14)(a) becomes

$$\frac{dP^\infty}{dt} \equiv 0 , \qquad \text{i.e.,} \quad P^\infty \equiv P_0 , \quad \text{a constant} ,$$

and $H(t) \equiv 0$ in (4.14)(b). In this case, we have the *constant pressure approximation* and *exothermic heat release acts as a source of specific volume in the fluid* - this is the intuitive content of the model in [29].

It is important that we summarize the conditions on the initial data which guarantee that the equations in (4.14) represent the valid low Mach number limit of the compressible combustion equations. We remind the reader that this is the case for the simpler incompressible limit in fluid dynamics studied extensively in Section 3 only when the initial data is slightly compressible, i.e.,

$$v^\lambda(x,0) = v_0^\infty(x) + \lambda^{-1} v_0^1(x) \qquad \text{with} \quad \operatorname{div} v_0^\infty = 0$$

$$p^\lambda(x,0) = P_0 + \lambda^{-2} p_0^1(x) .$$

The analogue of these conditions for the initial data for the compressible combustion equations in (4.1) is the requirement of *approximate chemical fluid belance* for the initial data. This condition is the requirement

$$\text{(4.16)}\quad \text{(a)}\qquad \begin{aligned} p^\lambda(x,0) &= P_0 + \lambda^{-2} p_0^1(x) + O(\lambda^{-3}) , & v^\lambda(x,0) &= v_0^\infty(x) + O(\lambda^{-1}) \\ T^\lambda(x,0) &= T_0^\infty(x) + O(\lambda^{-1}) , & z^\lambda(x,0) &= z_0^\infty(x) + O(\lambda^{-1}) \end{aligned}$$

where there exists a constant H_0 so that

$$\text{(4.16)}\quad \text{(b)}\qquad \operatorname{div} v_0^\infty = -H_0 + (P_0)^{-1}[\operatorname{div}(k\nabla T_0^\infty) + q_0 \frac{P_0}{T^\infty} W(z_0^\infty, T_0^\infty)] .$$

Of course, given arbitrary functions T_0^∞, z_0^∞, unique values of $\nabla\psi_0^\infty$ and H_0 are determined by (4.16)(b) - unlike ordinary incompressible fluid dynamics, the constraint in this case is nonlinear. In work to be published, the author [8] has recently verified that under the requirement of *approximate chemical fluid balance* on the initial data, for either all of space or the periodic case, *the system in* (4.14) is the low Mach number limit of the compressible combustion equations in (4.1). We remark here that the periodic case behaves like a bounded domain so this provides a justification of the formal asymptotics, and also a rigorous proof of the constant pressure approximation for all of space. Several new technical ideas beyond those sketched in Section 3 are needed for the proof. We also make the following

Remark. The equations of incompressible combustion imply conservation of mass, i.e., if $\rho^\infty = P^\infty/T^\infty$, then

$$\frac{D}{Dt}\rho^\infty + \rho^\infty \operatorname{div} v^\infty = 0 .$$

This fact is evident from our derivation. The system in (4.14) is the formal limit as $\lambda \to \infty$ of the system of compressible combustion equations in (4.1) which conserve mass - the skeptical reader can verify this directly from (4.14).

Incompressible Combustion in 1-D Lagrangian Coordinates

Here we examine the equations for zero Mach number combustion from (4.14) in Lagrangian mass coordinates and derive a new system of integro-differential reaction-diffusion equations. We assume that the domain $\overline{\Omega}$ is the interval $[0,L]$, and we introduce the change of variables

$$q(x,t) = \int_0^x \rho^\infty(s,t)\,ds \qquad (4.17)$$

$$t' = t\,.$$

If $M = \int_0^L \rho^\infty$, then q from (4.17) ranges over $0 \le q \le M$. As a consequence of our earlier remark, mass is conserved so that

$$\frac{\partial \rho^\infty}{\partial t} + \frac{\partial}{\partial x}(\rho^\infty v^\infty) = 0$$

and only easily computes from this equation and (4.17), the formulae

$$\frac{\partial}{\partial x} = \rho^\infty \frac{\partial}{\partial q}\,, \qquad \frac{\partial}{\partial t} + v^\infty \frac{\partial}{\partial x} \equiv \frac{D}{Dt} = \frac{\partial}{\partial t'}\,. \qquad (4.18)$$

Now, let us look at the equations for low Mach number combustion in (4.14) and specialize these equations to one space dimension. Since any function is a gradient in one space variable, we see that (4.14) is trivially satisfied and can be ignored. Next, we observe that it follows from (4.18) that *in a Lagrangian mass coordinate system, the velocity of the fluid does not have to be computed explicitly in equations* (4.14)(a), (d), (e) and these equations form a closed system. The velocity is uniquely determined from (4.14)(b) and can be found *afterwards* (if desired) once the solution of (4.14)(a), (d), (e) is determined in Lagrangian coordinates. Thus, the *equations for 1-D low Mach number combustion in Lagrangian coordinates become the much simpler system of integro-differential reaction diffusion equations*

(4.19)

$$\text{(a)} \quad \frac{dP^\infty}{dt'} = \frac{q_0}{\mathrm{Vol}(\Omega)} \int_0^M W(z^\infty,T^\infty)\,dq$$

$$\text{(b)} \quad \frac{\partial z^\infty}{\partial t'} = \frac{\partial}{\partial q}\left(\rho^2 d\, \frac{\partial z^\infty}{\partial q}\right) - W(z^\infty,T^\infty)\,, \qquad \frac{\partial z^\infty}{\partial q} = 0 \quad \text{at} \quad q = 0,M$$

$$\text{(c)} \quad \frac{\partial T^\infty}{\partial t'} = \frac{\gamma-1}{\gamma}\,\frac{1}{\rho}\,\frac{dP^\infty}{dt'} + \frac{\partial}{\partial q}\left(k\rho\, \frac{\partial T^\infty}{\partial q}\right) + q_0\, W(z^\infty,T^\infty)\,, \qquad \frac{\partial T^\infty}{\partial q} = 0 \quad \text{at} \quad q = 0,M$$

where

(4.19) (d) $\rho = P^{\infty}/T^{\infty}$.

(The double use of q for heat release and mass coordinate should not cause confusion.)

If we repeat the same derivation in an unbounded domain such as the right half-space $x > 0$, we have the constant pressure approximation and get the system of reaction-diffusion equations

$$\text{(4.20)}\quad \begin{aligned} &\text{(a)}\quad \frac{\partial z^{\infty}}{\partial t'} = \frac{\partial}{\partial q}\left(\rho^2 d\,\frac{\partial z^{\infty}}{\partial q}\right) - W(z^{\infty},T^{\infty}), && \frac{\partial z^{\infty}}{\partial q} = 0, \quad 0 < q, \quad t' > 0 \\ &\text{(b)}\quad \frac{\partial T^{\infty}}{\partial t'} = P_0\,\frac{\partial}{\partial q}\left(k(T^{\infty})^{-1}\,\frac{\partial}{\partial q} T^{\infty}\right) + q_0 W(z^{\infty},T^{\infty}), && \frac{\partial T^{\infty}}{\partial q} = 0, \quad 0 < q, \quad t' > 0 \end{aligned}$$

where $\rho = P_0(T^{\infty})^{-1}$. The system in (4.20) has nonlinear diffusion coefficients; except for this difference, the equations in (4.20) become the standard system of reaction-diffusion equations which many authors have used as simple model equations for combustion. However, we emphasize here that the previous derivations of reaction-diffusion equations occurred in Eulerian coordinates under a *constant density* approximation which is rather *unrealistic for combustion processes*. Here, we have derived a similar system in a completely different reference frame of Langrangian mass coordinates under *only the realistic assumptions* (see [21]) of a *constant pressure* approximation and approximate *chemical-fluid* balance for the initial data. This observation provides a quantitative reason for the success of reaction-diffusion equation models in predicting many qualitative effects in flame theory.

Finally, we conclude this section by posing the following problem:

> Elucidate the differences in solution behavior for the integro-differential reacting-diffusion system in (4.19) as compared with the more conventional reaction-diffusion equations.

Recently, some interesting theoretical work on a *scalar* integro-differential reaction diffusion equation has been done by Bebernes and Bressan [1]. This scalar equation arises from an asymptotic combustion model recently developed by Kassoy (see [12] for many references) in Eulerian coordinates under appropriate assumptions, including a constant density approximation. In fact, most aspects of this modeling process can be regarded as a special case of the simple equations derived in (4.19) - one notable exception is the inclusion of effects from thermal boundary layers in the approximations described in [11] (see [12]).

BIBLIOGRAPHY

[1] BEBERNES, J., and A. BRESSAN: "Thermal behavior for a confined reactive gas", J. Differential Equations 44 (1982), 118-133.

[2] CHORIN, A. J.: "The evolution of a turbulent vortex", Comm. Math. Phys. 83 (1982), 517-536.

[3] COURANT, R., and D. HILBERT: *Methods of Mathematical Physics*, Vol. II, Wiley-Interscience, New York, 1963.

[4] CRANDALL, M., and P. SOUGANIDIS: (in preparation).

[5] DOUGLIS, A.: "Some existence theorems for hyperbolic systems of partial differential equations in two independent variables", Comm. Pure Appl. Math. 5 (1952), 119-154.

[6] EBIN, D.: "The motion of slightly compressible fluids viewed as motion with a strong constraining force", Ann. Math. 150 (1977), 102-163.

[7] EBIN, D.: "Motion of slightly compressible fluids in a bounded domain. I", Comm. Pure Appl. Math. 35 (1982), 451-487.

[8] EMBID, P., and A. Majda: "Slightly compressible combustible fluids" (in preparation).

[9] GEYMONAT, G., and E. SANCHEZ-PALEWCIA: "On the vanishing viscosity limit for acoustic phenomena in a bounded region", Arch. Rational Mech. Anal. 75 (1981), 257-268.

[10] HARTMAN, P., and A. WINTER: "On hyperbolic differential equations", Amer. J. Math. 74 (1952), 834-864.

[11] KASSOY, D. R., and J. POLAND: "The thermal explosion confined by a constant temperature boundary: II - the extremely rapid transient", SIAM J. Appl. Math. 41 (1981), 231-246.

[12] KASSOY, D. R., and J. BEBERNES: "Gasdynamic aspects of thermal explosions", Trans. of Twenty-Seventh Conference of Army Math., pp. 687-706.

[13] KATO, T.: "Quasi-linear equations of evolution with applications to partial differential equations", Lecture Notes in Math. 448, Springer-Verlag (1975), 25-70.

[14] KATO, T.: "The Cauchy problem for quasi-linear symmetric hyperbolic systems", Arch. Rational Mech. Anal. 58 (1975), 181-205.

[15] KLAINERMAN, S., and A. MAJDA: "Singular limits of quasilinear hyperbolic systems with large parameters and the incompressible limit of compressible fluids", Comm. Pure Appl. Math. 34 (1981), 481-524.

[16] KLAINERMAN, S., and A. MAJDA: "Compressible and incompressible fluids", Comm. Pure Appl. Math. 35 (1982), 629-653.

[17] KLAINERMAN, S., and R. KOHN: "Compressible and incompressible elasticity" (in preparation).

[18] KREISS, H. O.: "Problems with different time scales for partial differential equations", Comm. Pure Appl. Math. 33 (1980), 399-441.

[19] LAX, P. D.: "Hyperbolic systems of conservation laws and the mathematical theory of shock waves", SIAM Reg. Conf. Lecture #11, Philadelphia, 1973.

[20] LIONS, J. L.: *Quelques Methodes de Résolution des Problèmes aux Limites Non Linéaires*, Dunod, Paris, 1969.

[21] MAJDA, A.: "Equations for low Mach number combustion", (to appear in Comb. Sci. and Tech.).

[22] MATKOWSKY, B. J., and G. I. SIVASHINSKY: "An asymptotic derivation of two models in flame theory associated with the constant density approximation", SIAM J. Appl. Math. 37 (1979), 686-699.

[23] MATSUMURA, A., and T. NISHIDA: "The initial value problem for the equations of motion of various and heat-conductive gases", J. Math. Kyoto Univ. 20 (1980), 67-104.

[24] MOSER, J.: "A rapidly convergent iteration method and nonlinear differential equations", Ann. Scuola Norm. Sup. Pisa 20 (1966), 265-315.

[25] TEMAM, R.: "Local existence of C^∞ solutions of the Euler equations of incompressible perfect fluids", in *Turbulence and the Navier-Stokes Equations*, Springer-Verlag, New York, 1976, 184-194.

[26] TEMAM, R.: *The Navier-Stokes Equations*, North Holland, Amsterdam, 1977.

[27] BEIRAO DA VEIGA, H.: "On the solutions in the large of the two-diemnsional flow of a non-viscous incompressible fluid" (preprint).

[28] CHORIN, A. J.: "A numerical method for solving incompressible viscous flow problems", J. Comput. Phys. 2 (1967), 12-26.

[29] GHONIEM, A.F., A.J. CHORIN, and A. K. OPPENHEIM: "Numerical modelling of turbulent flow in a combustion tunnel", Philos. Trans. Roy. Soc. London Ser. A (1981), 1103-1119.

[30] HALTINER, G. J., and R. T. WILLIAMS: *Numerical Weather Prediction and Dynamic Meteorology*, 2nd Edition, Wiley, New York, 1980.

[31] MAJDA, A.: "The existence of multi-dimensional shock fronts", Memoirs Amer. Math. Soc. (to appear 1983).

THE LINEAR TRANSPORT OPERATOR OF FLUID DYNAMICS

G. GEYMONAT
Politecnico di Torino
Istituto Matematico
Corso Duca degli Abruzzi 24
I-10129 Torino

P. LEYLAND
L.M.A.-C.N.R.S.
B.P. 71
F-13277 Marseille Cedex 9

1. Fluid dynamical problems, when posed in terms of a system of partial differential equations, lead automatically to the consideration of a first order operator of the form

$$A = \sum_{i=1}^{d} a_i(x,t) \frac{\partial}{\partial x_i}$$

the so-called transport or advection part of the fluid flow with velocity field $\{a_i\}$ $i=1,\dots d$ [1]. The associated linear evolution equation is studied here in some bounded region and is a typical example of the kind of problems arising from the hyperbolic terms in a more general context:

$$\frac{\partial u}{\partial t} + Au = f \qquad \text{in} \qquad \Omega \times (0,T), \qquad \Omega \subset \mathbb{R}^3 \tag{1}$$

with

$$u(t,x) = 0 \quad \text{for } x \text{ belonging to the part of } \partial\Omega \times (0,T) \text{ where the fluid enters } \Omega \times (0,T) \tag{2}$$

and

$$u(0,x) = u_o \tag{3}$$

In this talk we shall present some results largely inspired by Bardos 1 . Ω is taken to be a bounded open connected set of $\mathbb{R}^d$ with Lipschitz boundary $\partial\Omega$ in the sense of Nečas [2]; thus the exterior normal to $\partial\Omega$, $\vec{n} = \{n_i\}$, is defined for almost all $x \in \partial\Omega$. The boundary condition, (2), needs some comments. In fact we define

$$\partial\Omega_-(t) = \{x \in \partial\Omega \setminus \chi \; : \; \sum_{i=1}^{d} a_i(t,x)\, n_i(x) < 0\} \tag{4}$$

where χ is the subset of $\partial\Omega$ where $\vec{n}$ is not defined, and we set

$$\sum_{-} = \bigcup_{t\in(0,T)} \partial\Omega_{-}(t) \quad .$$

We remark that $\partial\Omega_{-}(t)$ and $\sum_{-}$ are measurable subsets of $\partial\Omega$ and $\partial\Omega \times (0,T)$.

The condition (2) will be taken in some integral mean sense (see below), for which we can add to $\sum_{-}$ any subset of $\partial\Omega \times (0,T)$ of zero surface measure.

Now, $\sum_{-} \cup \{\Omega \times \{t = 0\}\}$ represents the set where the flow enters Ω; reciprocally we may define the set where the flow exits Ω, by considering the set

$$\partial\Omega_{+}(t) = \{x \in \partial\Omega \setminus \chi : \sum_{i=1}^{d} a_i(t,x)\, n_i(x) > 0\} \ .$$

The set

$$\partial\Omega_{o} = \{x \in \partial\Omega \setminus \chi : \sum_{i=1}^{d} a_i(t,x)\, n_i(x) = 0\}$$

is of zero $(d-1)$-dimensional Lebesgue measure and thus never intervenes in the following.

We shall suppose throughout that the coefficients of A, $\{a_i\}_{i=1,\dots,d}$ are of $L^{\infty}(0,T;\, C^{0,1}(\bar{\Omega}))$ class.

Returning to the question over the boundary condition (2), in order to use local results the optimum solution is to work with an open set with the same measure as $\sum_{-}$; for this we assume throughout the following hypothesis

(B-):

$\left(\overline{\bigcup_{t\in(0,T)} \partial\Omega_{\pm}(t)}\right)^{o}$ (where closure and interior are taken with respect to $\partial\Omega \times (0,T)$), has the same d-dimensional surface Lebesgue measure as $\bigcup_{t\in(0,T)} \partial\Omega_{\pm}(t)$.

If $\partial\Omega$ is piecewise C^1, then (B-) is automatically satisfied [1], and the assumption (B-) seems natural also in the light of the following example, which seems pathological enough from a "practical" point of view.

Example

Let $\Omega = \{(x,y) \in \mathbb{R}^2 : 2 > y > \int_0^x g(t)dt = G(x),\ x \in\,]0,1[\}$ and $g : [0,1] \to [0,1]$ is a jump function discontinuous at all rational points of $[0,1]$. $G(x)$ is strictly convex, and Ω is of Lipschitz boundary, $\vec{n}$ is defined where g is continuous; since χ is dense in $[0,1]$, and is of zero one-dimensional measure, taking an operator A like $a_1 \frac{\partial}{\partial x} + a_2 \frac{\partial}{\partial y}$ then

$$\{x \in [0,1] \setminus \chi : a_1(x,G(x))g(x) - a_2(x,G(x)) < 0\} \subset \partial\Omega_-$$

for e.g. $a_2(x,G(x)) > a_1(x,G(x))$, then $[0,1] \setminus \chi \subset \partial\Omega_-$ and (B-) is satisfied.

Equally we introduce $\sum_+$ and thus we assume always hypothesis (B+).

From now on $\sum_\pm$ denotes the sets $\left(\overline{\bigcup_{t\in(0,T)} \partial\Omega_+(t)}\right)^{o}$

In the case of a stationary flow, $\{a_i\}_i$ is independent of time, and sets like $\partial\Omega_-(t)$ are simply denoted $\partial\Omega_-$, etc. Hypothesis (B) is still necessary and just becomes $\text{meas}_{d-1}(\partial\Omega_-) = \text{meas}_{d-1}(\overline{\partial\Omega_-})^{o}$. This case is given in §3, as we have more regularity on the solution of (1), (2), (3). In §2 we give a first estimation of solutions of (1), together with a condition for uniqueness. In §4 we obtain more regularity for the complete problem, $\{a_i\}$ depending on time, using the artificial viscosity method (parabolic regularisation).

2. We may consider (1), (2), (3) as the study of the first order operator

$$\mathcal{A} = \sum_{i=1}^{m} a_i(x) \frac{\partial}{\partial x_i}$$

then (1) becomes

$$\mathcal{A}\, u = f \tag{5}$$

for an open bounded subset $\theta \subset \mathbb{R}^m$. In this way for $m = d+1$ we consider $\theta \equiv Q \equiv \Omega \times (0,T)$ and $\{a_i(x)\}_{i=1,\dots,m} \equiv (1, a_i(t,x))_{i=1,\dots,d} = a(t,x)$ and

$$\mathcal{A} = \frac{\partial}{\partial t} + A(t)$$

for $m = d$, we consider $\mathcal{A} = A = \sum_{i=1}^{d} a_i(x) \frac{\partial}{\partial x_i}$ and we recuperate (1), (2), (3) for the time independent flow.

The sets $\sum_\pm$ are thus redefined in the obvious way and are denoted $\partial\theta_\pm$.

Let us consider the space

$$W(\mathcal{A}) = \{u \in L^2(\theta) : \mathcal{A}\, u \in L^2(\theta)\} .$$

Then we have

<u>Proposition 1.</u>

(i) $W(\mathcal{A})$ is a Hilbert space for the natural graph norm

(ii) $C^\infty(\overline{\theta})$ is dense in $W(\mathcal{A})$

(iii) The traces $u \to u\big|_{\Sigma_\pm}$ are linear and continuous from $W(\mathcal{A})$ to $L^2_{loc}(|\vec{a}\cdot\vec{n}|;\Sigma_\pm)$.

For details of the proof see [3].

We can thus define in particular the transport operator A (or $A(t)$) considered as an unbounded operator in $L^2(\Omega)$ by its domain

$$D(A) = \{u \in W(A) : u\big|_{\partial\Omega_-} = 0\} \tag{6}$$

and also we have

$$D(\mathcal{A}) = \{u \in W(\mathcal{A}) : u\big|_{\partial\Omega_-} = 0, \quad u(0,x) = 0\}.$$

From Proposition 1, $D(\mathcal{A})$ is well defined, it is a closed subspace of $W(\mathcal{A})$, dense in $L^2(\theta)$ and $\mathcal{A}$ is a closed operator in $L^2(\theta)$.

Let

$$\omega_o = \sup_{x \in \bar{\theta}} \text{ess} \left| \sum_{i=1}^{m} \frac{\partial}{\partial x_i} a_i(x) \right| .$$

The search for solutions of (5), with $u \in D(\mathcal{A})$, (and thus of (1), (2), (3)) is equivalent to the problem

$$(\mathcal{A} + \lambda \mathbb{1})\, u = f \tag{7}$$

by linearity, (by change of u to $e^{-\lambda t} u$).

We have in general the surjectivity property:

<u>Proposition 2</u>. For all $\lambda \in \mathbb{C}$, $\text{Re}\,\lambda > \frac{1}{2}\omega_o$, then

$$\text{Im}\,(\mathcal{A} + \lambda \mathbb{1}) = L^2(\theta) .$$

<u>Proof</u>. The proof can be done by elliptic regularisation; for this, we take

$$V = H^1_o(\theta)$$

and we consider the sesquilinear form $a_\varepsilon(u,v)$ on $V \times V$ given by

$$a_\varepsilon(u,v) = \varepsilon(\text{grad } u, \text{grad } v)_{L^2(\theta)} + (\mathcal{A}u, v)_{L^2(\theta)} + \lambda(u,v)_{L^2(\theta)} . \tag{8}$$

Integrating by parts we obtain

$$\text{Re } a_\varepsilon(u,u) \geq \inf\left(\varepsilon, \text{Re}(\lambda - \tfrac{1}{2}\,\omega_o)\right)\, |u|_V^2 . \tag{9}$$

So for all $\varepsilon > 0$, the regularised Dirichlet problem

$$\begin{aligned} & a_\varepsilon(u_\varepsilon, v) = (f,v) \qquad \text{for all} \quad v \in V \\ & u_\varepsilon \in V \end{aligned} \tag{10}$$

has a unique solution. Furthermore

$$\varepsilon |u_\varepsilon|^2_{H^1(\theta)} + \text{Re}\,(\lambda - \tfrac{1}{2}\,\omega_o)\, \|u_\varepsilon\|^2_{L^2(\theta)} \leq \|f\|_{L^2(\theta)} \|u_\varepsilon\|_{L^2(\theta)}$$

from which

$$\|u_\varepsilon\|_{L^2(\theta)} \leq c$$

and

$$\sqrt{2}\, \|\text{grad } u_\varepsilon\|_{L^2(\theta)} \leq c$$

(c is independent of ε), so there exists a convergent subsequence $\{u_\varepsilon\}$, which converges <u>weakly</u> in $L^2(\theta)$ to a certain u^*. From (10) then we have for all $v \in \mathcal{D}(\theta)$,

$$\sqrt{\varepsilon}\,(\sqrt{\varepsilon}\ \text{grad } u_\varepsilon, \text{grad } v)_{L^2(\theta)} + (u_\varepsilon, \mathcal{A}'v)_{L^2(\theta)} + \lambda(u_\varepsilon, v)_{L^2(\theta)} = (f,v)$$

where $\mathcal{A}'$ denotes the formal adjoint of $\mathcal{A}$); we can conclude by taking the limit $\varepsilon \to 0$ that $\mathcal{A}u^* \in L^2(\theta)$. Also

$$(\mathcal{A}(u_\varepsilon - u), v)_{L^2(\theta)} + \lambda((u_\varepsilon - u^*), v)_{L^2(\theta)} + \sqrt{\varepsilon}\,(\sqrt{\varepsilon}\ \text{grad } u_\varepsilon, \text{grad } v)_{L^2(\theta)} = 0$$

$$\text{for all} \quad v \in \mathcal{D}(\theta)$$

thus $\mathcal{A}u_\varepsilon$ tends to $\mathcal{A}u$ for $\varepsilon \to 0$ weakly in $L^2(\theta)$, and so $u_\varepsilon \to u^*$ in $W(\mathcal{A})$ weakly. Taking now the limit in (10), for every $v \in V$ we have that u^* verifies

$$((\mathcal{A} + \lambda \mathbb{1})u^*, v)_{L^2(\theta)} = (f,v)_{L^2(\theta)} .$$

Since $D(\mathcal{A})$ is a closed subspace of $W(\mathcal{A})$, (Proposition 1.iii), its strong closure coincides with its weak closure, thus we have also deduced that $u^* \in D(\ )$.

Q.E.D.

However with such a choice of V, we have no control on the convergence of the trace on $\sum_+$: indeed $u_\varepsilon\big|_{\sum_+}$ (which is zero) cannot converge to $u^*\big|_{\sum_+}$, which, by definition of $D(\ )$, has no reason to be zero; we have thus introduced a boundary layer effect near $\sum_+$; in order to avoid this, we could have used a regularisation via a mixed boundary value problem, as in 3 , taking as choice of V

$$V = \{v \in H^1(\Omega) : v\big|_{\sum_-} = 0\}$$

This boundary layer effect has often been examined in examples of regularised transport equations; in particular, the following two examples show this clearly.

Examples

(i) [4]. $-\varepsilon y''(x) + a(x)y'(x) + b(x)y(x) = 0$, $0 \le x \le 1$, and $y(0) = A$, $y(1) = B$. For coefficient a strictly positive, and continuous functions, this is a regularised Dirichlet transport equation for $\Omega = (0,1)$: $\partial\Omega_+ = \{1\}$, $\partial\Omega_- = \{0\}$; the solution has a boundary layer at $x = \{1\}$.

(ii) [5].

$$\left.\begin{aligned} -\varepsilon\,\Delta u(x,y) + \left(\frac{\partial}{\partial x} + b\,\frac{\partial}{\partial y}\right)u(x,y) &= 0 \quad \text{in} \quad \Omega \subset \mathbb{R}^2, \quad b > 0 \\ u(x,y) &= f \quad \text{on} \quad \partial\Omega \end{aligned}\right\}$$

which is a linear model associated to a nonlinear problem in chemical reactor theory with a transport term for the mass concentration. A boundary layer appears near $\partial\Omega_+$.

It could be questionable whether the presence of the boundary layer in the previous examples comes from the physically correct choice of boundary conditions; indeed, if we had taken mixed boundary conditions i.e. Dirichlet on $\sum_-$ and Neumann on the complementary, this would be more reasonable expecially for example (ii). The literature on this topic is very large, so we avoid to give any more references.

In order to obtain uniqueness of the solution of (7) we need to prove a positivity property for $\mathcal{A}$, i.e.

$$\text{for all } u \in D(\mathcal{A}), \quad \operatorname{Re}((\mathcal{A} + \lambda \mathbb{1})u,u)_{L^2(\theta)} \ge 0 \tag{11}$$

Now this is true for $u \in C^\infty(\bar\theta) \cap D(\mathcal{A})$ and $\operatorname{Re}\lambda > \frac{1}{2}\omega_o$. It is thus sufficient to prove that $C^{\infty}(\bar{\theta}) \cap D(\mathcal{A})$ is dense in $D(\mathcal{A})$, a nontrivial result for which we need the following additional hypothesis:

(H-) <u>The boundary of</u> $\sum_-$ <u>in</u> $\partial\theta$ <u>is of finite</u> (m-2)-<u>dimensional Hausdorff measure</u>.

In particular, if $\partial\theta$ is made up of a finite number of (m-2)-dimensional surfaces then this is true [1]. Details of the proof of this result can be found in [3].

3. We take a first $m = d$, then $\mathcal{A} = A$, $\theta = \Omega$ and we have

<u>Theorem 3</u>. $-(A + \frac{1}{2}\omega_o \mathbb{I})$ is an infinitesimal generator of a contraction semigroup in $L^2(\Omega)$. So, for all $f \in C([0,T]; L^2(\Omega))$, $u_o \in L^2(\Omega)$, there is one and only one solution of (1), (2), (3), and this solution $u \in C([0,T]; L^2(\Omega))$.
The proof follows from Proposition 2, (H-), (11), and thus the fact that A is in fact quasi-accretive [6].

4. Let us now take $m = d+1$, then $\mathcal{A} = \frac{\partial}{\partial t} + A(t)$ and $\theta = Q$, and $D(A(t))$ varies with time. From Proposition 2 we have that for $f \in L^2(Q) = L^2(0,T; L^2(\Omega))$ given, then there exists a solution of $L^2(Q)$ class, which is unique allowing for (H-). However, we can improve this result by using a parabolic regularisation technique for the problem (7), either without introducing any further hypotheses than those already introduced on the sets $\partial\Omega_{\pm}(t)$, $\sum_{\pm}$, or by imposing slightly more conditions and using explicitly the work of C. Baiocchi [7],[8]. The latter possibility is developped in [3], we will give here the former version.

Let

$$\tilde{V} = H^1_o(\Omega)$$

we redefine for fixed t, $t \in (0,T)$, the sesquilinear form $\tilde{a}_\varepsilon(t;u,v)$ on $\tilde{V} \times \tilde{V}$ as:

$$\tilde{a}_\varepsilon(t;u,v) = \varepsilon \Big(\sum_{i=1}^{d} \frac{\partial u}{\partial x_i} , \sum_{i=1}^{d} \frac{\partial v}{\partial x_i} \Big)_{L^2(\Omega)} + (A(t)\, u,v)_{L^2(\Omega)} + \lambda (u,v)_{L^2(\Omega)} \tag{12}$$

For $\{a_i\} \in L^{\infty}(0,T;C^{0,1}(\bar{\Omega}))$, then $t \to \tilde{a}_\varepsilon(t;u,v)$ is measurable on $(0,T)$, and bounded. Moreover, we have

$$\text{Re}\, \tilde{a}(t;u,v) \geq \inf\, (\varepsilon, \text{Re}(\lambda - \frac{1}{2}\omega_o)) \, |u|^2_V \tag{13}$$

Let

$$\Phi(0,T;\tilde{V}) = \{ \varphi \in L^2(0,T;H^1_o(\Omega)) \; : \; \varphi' \in L^2(0,T;L^2(\Omega) \quad \text{and} \quad \varphi(T) = 0 \} .$$

Then, for each $\varepsilon > 0$, we may consider the regularised parabolic problem for all $\varphi \in \Phi$, $f \in L^2(Q)$:

$$\int_o^T [\tilde{a}_\varepsilon(t;u_\varepsilon(t),\varphi_\varepsilon(t)) - (u_\varepsilon(t),\varphi'(t))_{L^2(\Omega)}]dt = \int_o^T (f(t),\, \varphi(t))_{L^2(\Omega)} dt$$

$$u_\varepsilon \in L^2(0,T;H^1_o(\Omega)) \quad ; \tag{14}$$

<u>Proposition 4</u>. Let $f \in L^2(Q)$, then (14) has a unique solution u_ε with

$$u_\varepsilon \in L^2(0,T;H^1_o(\Omega)) \cap H^{1/2}(0,T;L^2(\Omega)) \cap C([0,T];L^2(Q)) .$$

The proof is standard - see for example [9, §IV Theorem 2.1].

<u>Theorem 5</u>. Let $f \in L^2(Q)$, then (7) has a unique solution of class

$$H^{1/2}(0,T;L^2(\Omega)) \cap L^\infty(0,T;L^2(\Omega)) .$$

<u>Remark</u>. Let us point out that $u \in H^{1/2}(0,T;L^2(\Omega)) \cap L^\infty(0,T;L^2(\Omega))$ does not imply that $u \in C([0,T];L^2(\Omega))$.

<u>Proof</u>. <u>Existence</u>. Recalling the energy equality for (14) we find for $t \in (0,T)$, and all $\eta > 0$

$$||u_\varepsilon(t)||^2_{L^2(\Omega)} + 2\,\mathrm{Re} \int_o^t \tilde{a}_\varepsilon(s;u_\varepsilon,u_\varepsilon)ds \le \frac{1}{\eta} \int_o^t ||f||^2_{L^2(\Omega)} ds + \eta \int_o^t ||u_\varepsilon(s)||^2_{L^2(\Omega)} ds$$

Since we may adjust the constants with a suitable choice of λ (as before), then

$$||u_\varepsilon(t)||^2_{L^2(\Omega)} + \varepsilon \int_o^t |u_\varepsilon|^2_{\tilde{V}} ds + \lambda \int_o^t ||u_\varepsilon||^2_{L^2(\Omega)} ds \le c \int_o^T ||f(t)||^2_{L^2(\Omega)} ds .$$

We infer, as in Proposition 2, that $u_\varepsilon(t)$ remains in a bounded set of $L^\infty(0,T;L^2(\Omega))$, u_ε and $\sqrt{\varepsilon}\,\mathrm{grad}_x u_\varepsilon$ remain in a bounded set of $L^2(0,T;L^2(\Omega))$ with du_ε/dt remaining in a bounded set of $L^2(0,T;H^{-1}(\Omega)$. Thus by interpolation, u_ε belongs to a bounded set of $H^{1/2}(0,T;L^2(\Omega))$. Taking the limit $\varepsilon \to 0$, by extraction of a convergent subsequence $\{u_\varepsilon\}$ we find that

$u_\varepsilon \to u^*$ in $L^\infty(0,T;L^2(\Omega))$ for the weak-star topology,

$u_\varepsilon \to u^*$ in $H^{1/2}(0,T;L^2(\Omega))$ weakly.

As in the proof of Proposition 2, $u^* \in D(\mathcal{k})$. Yet again, we have introduced a boundary layer near $\sum_+$, as explained in § 2.

Uniqueness. Replacing $\partial\theta$ by ∂Q and m by $(d+1)$ in (H-), then uniqueness follows from (11).

Q.E.D.

5. Since $\tilde{V}$ is independent of t, then we may write (14) as

$$u'_\varepsilon(s) + \mathcal{A}_\varepsilon(s)\, u_\varepsilon(s) = f(s) \;, \qquad 0 \le s \le T$$

where $\tilde{a}_\varepsilon(s;u_\varepsilon(s),v) = (\mathcal{A}_\varepsilon(s)u_\varepsilon(s),v)_{L^2(\Omega)}$, $v \in \tilde{V}$.

Thus for $v \in \tilde{V}$ we have

$$<u'_\varepsilon(s),v>_{\tilde{V}',\tilde{V}} + \tilde{a}_\varepsilon(s;u_\varepsilon(s),v) = <f(s),v>_{\tilde{V}',\tilde{V}} \tag{15}$$

with $u_\varepsilon(s) \in L^2(0,T;\tilde{V})$, $u'_\varepsilon(s) \in L^2(0,T;\tilde{V}')$, $v(s) \equiv v$ for all s is in $L^2(0,T;\tilde{V})$. We can thus use the Green's formula (see [10, ch. I n. 5.3]), on the integrated version of (15) from 0 to t :

$$(u_\varepsilon(t),v)_{L^2(\Omega)} + \int_0^t \tilde{a}_\varepsilon(s;u_\varepsilon(s),v)ds = \int_0^t (f(s),v)_{L^2(\Omega)} ds \quad \text{for all } v \in \tilde{V} \tag{16}$$

We shall use this form of the (parabolic) regularised problem to prove the following further property of the solution of the linear transport equation.

Theorem 6. The solution u of (7) (and so, of (1)), is also of $BV(0,T;L^1(\Omega))$ class, where BV denotes the space of functions of bounded variation with values in $L^1(\Omega)$.

Proof. Consider the function for $\eta > 0$

$$\mathrm{sgn}_\eta(s) \begin{cases} 1 & \text{if} \quad s > \eta \\ \dfrac{s}{\eta} & \text{if} \quad |s| \le \eta \\ -1 & \text{if} \quad s < -\eta \end{cases}$$

If $v \in \tilde{V} = H^1_0(\Omega)$, then $\mathrm{sgn}_\eta(v) \in H^1_0(\Omega)$ because $\mathrm{sgn}_\eta(s)$ is a Lipschitz function in $\mathbb{R}$; so taking $v \in \mathrm{sgn}_\eta(u_\varepsilon(t))$ in (16), we find

$$\int_{\Omega_\eta} |u_\varepsilon(t)|ds + \frac{1}{\eta} \int_{\Omega\backslash\Omega_\eta} |u_\varepsilon(t)|^2 dx + \int_0^t \tilde{a}_\varepsilon(t;u_\varepsilon(s),\text{sgn}_\eta(u_\varepsilon(s)))\, ds$$

$$= \int_0^t \int_{\Omega_\eta} |f(s)|dx\, ds + \frac{1}{\eta} \int_0^t \int_{\Omega\backslash\Omega_\eta} f(s)\, \overline{u_\varepsilon(s)}\, dx\, ds \tag{17}$$

where $\Omega_\eta = \{x \in \Omega : |u_\varepsilon(t,x)| > \eta\}$

$\Omega\backslash\Omega_\eta = \{x \in \Omega : |u_\varepsilon(t,x)| \le \eta\}$

Recalling (13), then (17) gives

$$\int_{\Omega_\eta} |u_\varepsilon(t)|dx \le \int_0^t \int_\Omega |f(s)|dx\, ds$$

thus for $\eta \to 0$, $\|u_\varepsilon(t)\|_{L^1(\Omega)} \le \int_0^t \|f(s)\|_{L^1(\Omega)} ds$, which means that $u_\varepsilon(t)$ belongs to a bounded set of $BV(0,T;L^1(\Omega))$. When $\varepsilon \to 0$ in the regularised problem, then $u_\varepsilon \to u^*$ solution of (7) is not only in $L^\infty(0,T;L^2(\Omega)) \cap H^{1/2}(L^2(\Omega))$ class as in Theorem 5, but also $u^* \in BV(0,T;L^1(\Omega))$.

Q.E.D.

One application of the above results is in the study of the system governing the transport and propagation of a linear acoustic (i.e. compressible) perturbation in a compressible (or incompressible) flow: written symbolically as

$$\frac{dw}{dt} + Aw + Bw = f$$

where A is the transport-advection operator studies above, and B is a Stokes type operator. This system is obtained by linearisation of the complete nonlinear system of equations of a compressible newtonian fluid with respect to a compressible perturbation w. The above techniques are taken over to the operator $(A+B)$, and we obtain the same kind of results (see [6]) for the now vectorial solution w. In fact, since B is essentially an elliptic operator, the main difficulty comes from the mass density, ρ, equation:

$$\frac{\partial\rho}{\partial t} + \sum_{i=1}^{d} a_i(t)\frac{\partial\rho}{\partial x_i} + \rho \sum_{i=1}^{d} \frac{\partial}{\partial x_i} a_i(t) = f_\rho - \alpha(a_i) \sum_{i=1}^{d} \frac{\partial u_i}{\partial x_i} - \beta(a_i) \tag{18}$$

written here for a barotropic fluid, where $\{u_i\}$ is the velocity field of the perturbation, $w = (u_i,\rho)_{i=1,\dots,d}$, α and β are coefficients depending on the flow variables. Whereas the other equations of the system have viscosity regularising terms.

(18) is then treated as a transport evolution problem, and we apply the techniques employed here to obtain existence, uniqueness and regularity results (see [3]).

REFERENCES

[1] C. BARDOS, Problèmes aux limites pour les équations aux dérivées partielles du premier ordre. Ann. Scient. Ec. Norm. Sup., 4e Série, t. 3 (1970), 185-233.

[2] J. NEČAS, Méthodes Directes en Théorie des Equations Elliptiques. Edit. Masson, Paris, 1967.

[3] G. GEYMONAT, P. LEYLAND, Transport and Propagation of a linear acoustic perturbation through a flow in a bounded region. To appear.

[4] C. BENDER, S. ORSZAG, Advanced Mathematical Methods for Scientists and Engineers. Ed. McGraw-Hill, 1975.

[5] D. S. COHEN, Perturbation Theory, in Modern modelling of continuum phenomena. Lectures in Appl. Math., 16, AMS, 61-108 (1977).

[6] T. KATO, Perturbation Theory for Linear Operators. Second Ed., Springer-Verlag, 1976.

[7] C. BAIOCCHI, Regolarità e unicità della soluzione di una equazione differenziale astratta, Rend. Sem. Mat. Padova, XXXV (1965), 380-417.

[8] C. BAIOCCHI, Sul problema misto per l'equazione parabolica del tipo del calore. Rend. Sem. Mat. Padova, XXXIV (1966), 80-121.

[9] J. L. LIONS, Equations Différentielles Opérationnelles et Problèmes aux Limites. Springer Verlag (1961).

[10] J. L. LIONS, in "Equazioni differenziali astratte". C.I.M.E. 1° Ciclo 1963, Varenna.

LIMITES DES EQUATIONS D'UN FLUIDE COMPRESSIBLE LORSQUE LA COMPRESSIBILITE TEND VERS ZERO

A. LAGHA-BENABDALLAH
Université des Sciences et de la Technologie d'Alger
Institut de Mathématiques

Dans ce travail on s'intéresse aux solutions des équations de la dynamique des gaz :

$$\rho(\frac{\partial u}{\partial t} + (u.\nabla)u) - \nu.\Delta u = - \nabla p \quad (1)$$

$$\frac{\partial \phi}{\partial t} + \nabla.(\phi u) = 0 \quad (2)$$

$$u(x,0) = w_o(x) \quad (3)$$

$$\rho(x,0) = \overline{\rho}_o(x) \quad (4)$$

où les inconnues sont la vitesse u et la densité ρ . La pression p est une fonction donnée de ρ .

L'existence locale en temps a fait l'objet de plusieurs travaux, en particuliers lorsque les équations (1) et (2) sont couplées avec une équation d'évolution pour la température θ . Parmi ces travaux on peut citer ceux de Nash [14] , Itaya [3], Tani [16].

L'existence globale en temps a fait l'objet de travaux plus récents, très différents les uns des autres. Parmi eux on peut noter Kanel [7], Itaya [4,5], Kazhykov et Setuklin [10]. Mais chacun des auteurs n'a traité que des cas particuliers (tels Kazhykov et Selukhin qui ont résolu le cas d'un gaz parfait où l'espace Ω est réduit à l'intervalle $[0,1]$). Le premier travail, assez général a été celui de Nishida et Matsumura [15]. Ces derniers ont établi l'existence globale de la solution (u,ρ,θ) des équations (1), (2), (3), (4) employées avec une équation d'évolution sur la température θ . Dans leur travail Nishida et Matsumura traitent le cas de la dimension trois où le gaz est considéré comme parfait et polytropique. Pour obtenir l'existence de la solution (u,ρ,θ) sur

$\mathbb{R}^+ \times \mathbb{R}^3$, ils sont amenés à imposer des contraintes sur la donnée initiale $(u_o, \overline{\rho}_o, \theta_o)$, du type suivant :

$$||u_o||_3^2 + ||\overline{\rho}_o||_3^2 + ||\theta_o||_3^2 \leqslant A$$

où les normes sont prises dans H^3 $(\mathbb{R}^2)$.

Récemment Klainerman et Majda [18] ont prouvé l'existence locale de la solution (u,ρ) des équations (1), (2), (3), (4) et de celle des équations d'Euler compressibles (correspondant à ν égal à zéro) dans le cas où Ω est le tore T^n de l'espace $\mathbb{R}^n$. Les seules hypothèses sur le donnée initiale sont d'être "incompressibles" :

$$\overline{\rho}_o \text{ constant} \quad , \quad \nabla . u_o = 0$$

En fait Klainerman et Majda se sont intéressés au problème de la convergence de la solution (u,ρ) vers celle des équations de Navier-Stokes incompressibles :

$$\frac{\partial u}{\partial t} + (u.\nabla)u - \frac{\nu}{\rho_o} \Delta u = - \frac{\nabla p}{\rho_o} \tag{5}$$

$$\nabla . u = 0 \tag{6}$$

$$u(x,0) = u_o(x) \tag{7}$$

Cette convergence ayant lieu lorsque la compressibilité tend vers zéro.

Pour notre part nous nous intéressons également aux solutions du système compressible (1), (2), (3), (4) lorsque la compressibilité tend vers zéro. En effet nous savons que dans le nature tout fluide est plus ou moins compressible, et dire que tel fluide est incompressible c'est en fait, faire une approximation (dans le cas de compressibilités très petites). Aussi est-il important de pouvoir justifier de telles approximations. Notre travail s'est beaucoup inspiré des techniques utilisées par Nishida et Matsumura [15], le résultat obtenu est global en temps mais n'est valable que, comme dans [15] , pour des données initiales petites. Le résultat rejoint celui de Klainerman et Majda [18], bien que plus général car global en temps. Dans [18] les techniques utilisées sont totalement différentes des notres. En utilisant la théorie des espaces de Sobolev avec poids et anisotropes (où le rôle de poinds est joué par la compressibilité), les auteurs établissent la convergence de la solution (u,p) des équations (1), (2), (3), (4) vers celle des équations (5), (6), (7) dans $[0,T_1] \times T^n$. Par les techniques qu'ils introduisent, Klainerman et Majda obtiennent un résultat qui s'applique à certains types de systèmes quasi linéaires, où des hypothèses sont imposées à la structure non linéaire du système (en particulier de symétrisabilité) et à la donnée initiale (qui généralisent celles imposées dans le cas du système de la dynamique des gaz, c'est-à-dire "d'incompressibilité").

Considérons ρ comme fonction de p. La compressibilité est exprimée par la variation de ρ par rapport à p. Un fluide incompressible est un fluide à densité constante.

En effectuant le développement limité de $p(\rho)$ à l'ordre un au voisinage de ρ_o, on obtient :

$$p(\rho) \cong p(\rho_o) + (\rho\ \rho_o)\frac{\partial p}{\partial \rho}(\rho_o)$$

où ρ_o est une constante arbitraire.

On pose alors :

$$\varepsilon = \left(\frac{\partial p}{\partial \rho}(\rho_o)\right)^{-1}$$

On appelle alors p'_ε l'expression $(p_\varepsilon(\rho_\varepsilon) - p_\varepsilon(\rho_o))$ et on identifie p_ε à p'_ε.

Ce qui donne :

$$\rho_\varepsilon(p_\varepsilon) = \rho_o + \varepsilon\, p_\varepsilon \tag{8}$$

Ce qui, introduit dans le système (1), (2), (3), (4), donne :

$$\rho_\varepsilon(p_\varepsilon)\left(\frac{\partial u_\varepsilon}{\partial t} + (u_\varepsilon.\nabla)u_\varepsilon\right) - \nu\,\Delta\, u_\varepsilon = -\nabla p_\varepsilon \tag{9}$$

$$\varepsilon\,\frac{\partial p_\varepsilon}{\partial t} + \varepsilon\, u_\varepsilon.\nabla p_\varepsilon + \rho_\varepsilon(p_\varepsilon)(\nabla.u_\varepsilon) = 0 \tag{10}$$

$$u_\varepsilon(x,0) = u_o(x) \tag{11}$$

$$p_\varepsilon(x,0) = p_o(x) \tag{12}$$

On montre alors, que pour des données initiales petites, vérifiant :

$$\operatorname{div} u_o = 0 \tag{13}$$

$$p_\varepsilon(x,0) = p_o(x) \qquad \forall\varepsilon > 0 \tag{14}$$

la solution $(u_\varepsilon, p_\varepsilon)$ de (9), (10), (11), (12) converge dans un sens convenable, lorsque ε tend vers zéro, vers (u,p) solution de (5), (6), (7).

Notre travail est divisé en deux parties. Dans la première, on montre l'existence globale de $(u_\varepsilon, p_\varepsilon)$ vérifiant des estimations indépendantes de ε, dans la seconde, on passe à la limite sur ε. Mais le travail essentiel est situé dans le première partie où nous montrons les estimations indépendantes de ε que vérifient u_ε et $\sqrt{\varepsilon}\, p_\varepsilon$.

Notre méthode consiste à utiliser à la fois les techniques introduites dans [15] et celles développées par R. Temam [17] dans le chapitre intitulé "la compressibilité artificielle" (ce qui lui donne un nouveau théorème d'existence pour les équations limites). En effet, si on applique directement le travail de Nishida et Malsumura, on n'arrive pas à obtenir les estimations indépendantes de ε, et si on transcrit les techniques de Temam, on n'obtient aucune estimation (ce qui est dû au terme $\rho_\varepsilon(u_\varepsilon.\nabla)u_\varepsilon$ qui ne figure pas dans les équations traitées par Temam [17]).

Dans la première partie (I), nous cherchons à montrer les estimations sur $(u_\varepsilon, p_\varepsilon)$, pour y arriver nous procédons exactement comme dans [15]. C'est-à-dire que, dans le paragraphe I.1., nous établissons l'existence d'un temps T_ε (dans toute la première partie, le travail se fait à ε fixé) tel que sur $[0,T_\varepsilon] \times \mathbb{R}^2$ les équations (9), (10), (11), (12) admettent une unique solution dans $H^3(\mathbb{R}^2) \times H^3(\mathbb{R}^2)$ qui vérifie une estimation du type :

$$E_\varepsilon(t) = \|u_\varepsilon(t,.)\|_2^2 + \varepsilon|D^3u_\varepsilon(t,.)|^2 + \varepsilon\,\|p_\varepsilon(t,.)\|_2^2 + \varepsilon^2|D^3p_\varepsilon(t,.)|^2 \leqslant 4\,E_o \tag{15}$$

où

$$E_o = ||u_o||_2^2 + \varepsilon|D^3u_o|^2 + \varepsilon\,||p_o||_2^2 + \varepsilon^2\,|D^3p_o|^2 \tag{16}$$

Dans le paragraphe I.2., on montre alors que si la solution $(u_\varepsilon, p_\varepsilon)$ existe sur $[0,T] \times \mathbb{R}^2$, où T est quelconque, et vérifie :

$$\underset{t\varepsilon[0,T]}{\mathrm{Max}}\ E_\varepsilon(t) \leqslant A \tag{17}$$

où A est une constante arbitraire, alors il existe C_o, indépendant de ε et de T, tel que $(u_\varepsilon, p_\varepsilon)$ vérifie :

$$\underset{t\varepsilon[0,T]}{\mathrm{Max}}\ E_\varepsilon(t) \leqslant C_o\,E_o \tag{18}$$

Pour en déduire l'existence globale et les estimations cherchées, il suffit, et c'est l'objet du paragraphe I.3. de conclure. En effet, tenant compte du paragraphe I.1. en imposant à E_o de vérifier :

$$E_o \leqslant \frac{A}{4} \tag{19}$$

les hypothèses du paragraphe I.2. sont vérifiées sur $[0,T_\varepsilon]$ et donc on a :

$$E_\varepsilon(t) \leqslant C_o\,E_o \qquad \forall t\ \varepsilon\ [0,T_\varepsilon] \tag{20}$$

En choisissant T_ε comme instant initial, on déduit comme dans le paragraphe I.1. l'existence de $(u_\varepsilon, p_\varepsilon)$ sur $[T_\varepsilon, 2T_\varepsilon]$ tel que :

$$E_\varepsilon(t) \leqslant 4\,E_\varepsilon(T_\varepsilon) \qquad \forall t\ \varepsilon\ [T_\varepsilon, 2T_\varepsilon]$$

Donc, d'après (20) :

$$E_\varepsilon(t) \leqslant 4\,C_o\,E_o \qquad \forall t\ \varepsilon\ [T_\varepsilon, 2T_\varepsilon] \tag{21}$$

Il suffit alors de choisir :

$$E_o \leqslant \frac{A}{4C_o} \tag{22}$$

et de (17) et (21), on en déduit :

$$E_\varepsilon(t) \leqslant A \qquad \forall t\ \varepsilon\ [0, 2T_\varepsilon]$$

Ce qui nous permet d'appliquer de nouveau les résultats du paragraphe I.2. à $[0, 2T_\varepsilon]$ et ainsi de suite. Ceci nous permet d'obtenir l'existence globale de $(u_\varepsilon,p_\varepsilon)$, vérifiant :

$$E_\varepsilon(t) \leqslant C_o E_o \qquad \forall t \in \mathbb{R}^+$$

dès que :

$$E_o \leqslant \min\left(\frac{A}{4}, \frac{A}{4C_o}\right) \tag{23}$$

Pour passer à la limite sur ε , il reste à trouver une estimation sur la dérivée en temps de $(u_\varepsilon, p_\varepsilon)$, ce qui s'obtient par des procédés classiques (cf. paragraphe II).

C'est dans le paragraphe I.2. que se situe l'essentiel du travail. On commence par montrer une inégalité d'énergie, puis on transforme le système (9), (10), (11), (12) en introduisant l'inconnue :

$$v_\varepsilon = \rho_\varepsilon . u_\varepsilon \tag{24}$$

qui, physiquement, représente la densité du flux de masse. On obtient alors le système suivant :

$$\frac{\partial v_\varepsilon}{\partial t} + (v_\varepsilon.\nabla).u_\varepsilon + u_\varepsilon \nabla.v_\varepsilon - \nu\Delta u_\varepsilon = -\nabla p_\varepsilon \tag{25}$$

$$\varepsilon \frac{\partial p_\varepsilon}{\partial t} + \text{div } v_\varepsilon = 0 \tag{26}$$

$$v_\varepsilon(x,0) = (\rho_o \; u_o)(x) \tag{27}$$

$$p_\varepsilon(x,0) = p_o(x) \tag{28}$$

d'inconnue v_ε et p_ε .

Ce système se rapproche de celui introduit par Temam, mais diffère par le terme :

$$(v_\varepsilon.\nabla)u_\varepsilon + u_\varepsilon \nabla.v_\varepsilon - \nu\Delta u_\varepsilon$$

Dans le système étudié par Temam [17], la quantité u_ε n'apparaît pas.

Le nouveau système (25), (26), (27), (28) permet d'obtenir les estimations indépendantes de ε . En effet, en procédant aux opérations classiques dérivation, intégration par parties le terme génant, ∇p_ε , disparait avec celui de (26) : $\nabla.v_\varepsilon$; ce qui se passe également dans les équations traitées par Temam [17]. Mais l'inconvénient est que les estimations obtenues pour :

$$\|u_\varepsilon(t,.)\|_2^2 + \varepsilon\,\|p_\varepsilon(t,.)\|_2^2$$

sont fonction de :

$$\varepsilon\,|||\,D^3 p_\varepsilon|||'_{L^2([0,T]\ ;\ L^2(\mathbb{R}^2))}$$

C'est ce qui nous amène à, de nouveau, appliquer les techniques de Nishida, Matsumura [15], pour chercher les estimations sur les dérivées troisièmes de la solution $(u_\varepsilon, p_\varepsilon)$ du système (9), (10), (11), (12). C'est ce qui nous introduit la dépendance en ε pour ces quantités ; et l'estimation obtenue est du type :

$$\|u_\varepsilon(t,.)\|_2^2 + \varepsilon|D^3 u_\varepsilon(t,.)|^2 + \varepsilon\,\|p_\varepsilon(t,.)\|_2^2 + \varepsilon|D^3 p_\varepsilon(t,.)|^2 \leqslant C_o\ E_o$$

Et donc :

$$\|u_\varepsilon(t,.)\|_2^2 + \varepsilon\,\|p_\varepsilon(t,.)\|_2^2 \leqslant C_o\ E_o \leqslant \text{constante}$$

ce qui donne une estimation indépendante de ε suffisante pour le passage à la limite sur ε.

Cet article contient essentiellement les résultats d'une thèse de 3ème cycle, présentée à l'Université de Paris-Nord en 1979 , et je tiens à exprimer tous mes remerciements à C. Bardos, Professeur à l'Université de Paris-Nord et à S. Bénachour, Professeur à l'Université Houari Boumedienne, sans qui ce travail n'aurait pu être fait.

I.1. Généralité et existence locale

On considérera le cas où Ω est l'espace $\mathbb{R}^2$

$$x = (x_1, x_2)$$

désignera la variable d'espace.

On notera :

$$\nabla u = \left(\frac{\partial u_j}{\partial x_j}\right)_{i,j=1,2}$$

$$(u.\nabla)u = \sum_{i=1}^{i=2} u_i \frac{\partial u}{\partial x_i}$$

$$\nabla^k u = \left(\left(\frac{\partial}{\partial x_i}\right)^\alpha u_i \ ; \ |\alpha| = k \quad i = 1,2, \quad \alpha = (\alpha_i,\alpha_2)\right)$$

$$\partial_j u = \frac{\partial u}{\partial x_j}$$

Les espaces $L^p(\mathbb{R}^2)$ et $H^s(\mathbb{R}^2)$, où p parcourt l'ensemble N et s est un nombre réel, désigneront les espaces classiques décrits dans [13] ; les normes dans ces espaces seront notées :

$$||u||_{L^p(\mathbb{R}^2)} \quad \text{et} \quad ||u||_s .$$

De même on utilisera les espace de $L^p([0,T] \ ; H^s(\mathbb{R}^3))$ munis de la norme :

$$|||u||| = \left(\int_0^T ||u(\tau,.)||_s^p d\tau\right)^{1/p} .$$

Tout au long de notre travail nous aurons à appliquer le lemme suivant (démontrer par exemple dans [11]) :

Lemme 1 : *a) Il existe une constante b_1 positive telle que :*

$$||v||_{L^4(\mathbb{R}^2)} \leq b_1 |v|^{1/2} \, |\nabla v|^{1/2} \quad \forall v \in L^4(\mathbb{R}^2) \tag{29}$$

(où $|v|$ désigne la norme de v dans $L^2(\mathbb{R}^2)$).

b) Il existe une constante positive b_2 telle que :

$$\|v\|_{L^\infty(\mathbb{R}^2)} \leqslant b_2 \|v\|_2 \quad \forall v \in L^\infty(\mathbb{R}^2) \tag{30}$$

c) $H^2(\mathbb{R}^2)$ *est une algèbre.*

L'existence d'une solution régulière pour un temps petit résulte du théorème suivant :

Théorème 1 : *Pour tout couple* (u_o,p_o) *appartient à l'espace* $(H^3(\mathbb{R}^2))^3$ *il existe trois constante* T_ε, ε_o, ν_o *telles que l'on ait les assertions suivantes :*

i) ε_o *et* ν_o *sont indéoendants de* ε

ii) Sur l'intervalle $[0,T_\varepsilon]$, *le système (9), (10), (11), (12) avec donnée initiale* (u_o,p_o) *admet une unique solution appartenant à l'espace* $(L^\infty([0,T_\varepsilon]\ ;\ H^3(\mathbb{R}^2)))^3$. *De plus* ∇u_ε *appartient à* $(L^2([0,T_\varepsilon]\ ;\ H^3(\mathbb{R}^2)))^2$.

iii) L'énergie $E_\varepsilon(t)$ *vérifie l'inégalité suivante :*

$$E_\varepsilon(t) + \nu_o \int_0^t \|\nabla u_\varepsilon(\tau,\cdot)\|_2^2 d\tau + \nu_o\ \varepsilon \int_0^t |D^4 u_\varepsilon(\tau,\cdot)|^2 d\tau \leqslant 4E_o \tag{31}$$

Démonstration

On reprend le travail de Nishida et Matsumura [15]. Dans une première étape on construit, par un procédé de récurrence une suite $(u_\varepsilon^n, p_\varepsilon^n)_{n\in\mathbb{N}}$ et dans la seconde étape, on montre l'existence de T_ε tel que pour tout n entier, $(u_\varepsilon^n, p_\varepsilon^n)$ vérifie (31) sur $[0,T_\varepsilon]$, pour ε assez petit $(\varepsilon \leqslant \varepsilon_o)$.

On montre alors l'existence de $(u_\varepsilon, p_\varepsilon)$ sur $[0,T_\varepsilon] \times \mathbb{R}^2$, comme limite de la suite $(u_\varepsilon^n, p_\varepsilon^n)_{n\in\mathbb{N}}$ lorsque n tend vers l'infini ; $(u_\varepsilon, p_\varepsilon)$ vérifie (31) à la limite .

I.2. Premières estimations à priori indépendantes de ε

En supposant l'existence de la solution $(u_\varepsilon, p_\varepsilon)$, et sous certaines hypothèses, on montre des estimations sur $(u_\varepsilon, \sqrt{\varepsilon}\ p_\varepsilon)$, indépendantes de ε. Celles-ci, comme cela a été expliqué dans l'introduction permettent à la fois de montrer l'existence globale de la solution et de passer à la limite sur ε.

On sait déjà que les équations (9), (10), (11) et (12) admettent dans

$[0,T_\varepsilon| \times \mathbb{R}^2$ une unique solution où T_ε est assez petit par rapport à la quantité.

$$E_3 = \|u_\varepsilon(0,\cdot)\|_2^2 + \varepsilon|D^2 u_\varepsilon(0,\cdot)|^2 + \varepsilon\|p_\varepsilon(0,\cdot)\|_2^2 + \varepsilon^2|D^3\ p_2(0,\cdot)|^2 \ .$$

De plus sur cet intervalle le quantité

$$E_\varepsilon(t) = \|u_\varepsilon(t,\cdot)\|_2^2 + \varepsilon|D^3 u_\varepsilon(t,\cdot)|^2 + \varepsilon\ \|p_\varepsilon(t,\cdot)\|_2^2 + \varepsilon^2|D^3 p_\varepsilon(t,\cdot)|$$

reste finie.

On introduit pour T et A positifs (T peut être petit et A grand, mais ce qui sera important c'est que toutes les estimations seront indépendantes de T et de ε) et on dira qu'une fonction $(u_\varepsilon, p_\varepsilon)$ satisfait à l'hypothèse H(A,T) (ou plus brièvement) l'hypothèse H si elle est solution sur l'intervalle $[0,T]$ des équations (9) - (12) et si sur l'intervalle $[0,T]$ elle vérifie l'estimation

$$\underset{t\in[0,T]}{\text{Max}}\ E_\varepsilon(t) \leq A \tag{32}$$

On supposera désormais que l'on a choisi A et T pour que cette hypothèse soit satisfaite et on se propose d'établir des estimations dépendant de A mais indépendantes de T et de ε .

Le première relation est du type énergie et c'est l'objet du :

<u>*Théorème 2*</u> *: Il existe trois constantes C_1, ν_1, ε_1 (dépendant de A), mais indépendantes de ε et de T telles que pour tout $t \in [0,T]$ et $\varepsilon \leq \varepsilon_1$, on ait :*

$$|u_\varepsilon(t,\cdot)|^2 + \varepsilon|p_\varepsilon(t,\cdot)|^2 + \nu_1\int_0^t |\nabla u_\varepsilon(\tau,\cdot)|^2\, d\tau \leq C_1\, E_0 \tag{33}$$

<u>Démonstration</u>

Pour $t \in [0,T]$ on obtient en faisant le produit scalaire de (9) avec u_ε :

$$\int_{\mathbb{R}^2} \rho_\varepsilon \frac{\partial u_\varepsilon}{\partial t} . u_\varepsilon\, dx + \int_{\mathbb{R}^2} \rho_\varepsilon\ (u_\varepsilon.\nabla)u_\varepsilon.u_\varepsilon\, dx + \nu|\nabla u_\varepsilon|^2 = -\int_{\mathbb{R}^2} \nabla p_\varepsilon.u_\varepsilon\, dx \tag{34}$$

on a :

$$\int_{\mathbb{R}^2} \nabla p_\varepsilon.u_\varepsilon dx = -\int_{\mathbb{R}^2} p_\varepsilon \nabla.u_\varepsilon dx \quad \text{et} \quad \int_{\mathbb{R}^2} \rho_\varepsilon \frac{\partial u_\varepsilon}{\partial t}.u_\varepsilon dx = \frac{1}{2}\int_{\mathbb{R}^2} \rho_\varepsilon \frac{d}{dt}|u_\varepsilon|^2\, dx$$

On en déduit l'égalité :

$$\int_{\mathbb{R}^2} \rho_\varepsilon \frac{d}{dt} |u_\varepsilon|^2 dx + \int_{\mathbb{R}^2} \rho_\varepsilon (u_\varepsilon . \nabla) u_\varepsilon . \nabla u_\varepsilon dx + \nu |\nabla u_\varepsilon|^2 = \int_{\mathbb{R}^2} p_\varepsilon \nabla . u_\varepsilon dx \tag{35}$$

ii) On fait le produit scalaire dans $L^2(\mathbb{R}^2)$ de (10) avec $(\frac{1}{2} u_\varepsilon . u_\varepsilon)$:

$$\int_{\mathbb{R}^2} \frac{|u_\varepsilon|^2}{2} \frac{\partial}{\partial t} \rho_\varepsilon dx + \frac{1}{2} \int_{\mathbb{R}^2} \nabla . (\rho_\varepsilon u_\varepsilon) u_\varepsilon . u_\varepsilon dx = 0 \tag{36}$$

En additionnant (35) et (36) on obtient :

$$\frac{1}{2} \frac{d}{dt} |\sqrt{\rho_\varepsilon}\, u_\varepsilon|^2 + \nu |\nabla u_\varepsilon|^2 = \int_{\mathbb{R}^2} p_\varepsilon \nabla . u_\varepsilon dx \tag{37}$$

où la positivité de ρ_ε sera justifiée ultérieurement. En effet :

$$\int_{\mathbb{R}^2} \rho_\varepsilon \frac{\partial}{\partial t} \frac{|u_\varepsilon|^2}{2} dx + \int_{\mathbb{R}^2} \frac{u_\varepsilon . u_\varepsilon}{2} \frac{\partial}{\partial t} \rho_\varepsilon dx = \frac{1}{2} \frac{d}{dt} |\sqrt{\rho_\varepsilon} . u_\varepsilon|^2$$

et

$$\int_{\mathbb{R}^2} \rho_\varepsilon (u_\varepsilon . \nabla) u_\varepsilon . u_\varepsilon dx + \frac{1}{2} \int_{\mathbb{R}^2} \nabla . (\rho_\varepsilon u_\varepsilon) . u_\varepsilon u_\varepsilon \, dx = 0 .$$

iii) On désigne par W la primitive de $\dfrac{p_\varepsilon(\rho_\varepsilon)}{\rho_\varepsilon^2}$ nulle pour ρ_ε égal à ρ_0

$$W(\rho_\varepsilon) = \int_{\rho_0}^{\rho_\varepsilon} \frac{p_\varepsilon(s)}{s^2} ds \tag{38}$$

qui est une fonction positive de ρ_ε.

En faisant le produit scalaire dans $L^2(\mathbb{R}^2)$ de (10) avec :

$$(\rho_\varepsilon \frac{\partial W}{\partial \rho_\varepsilon} (\rho_\varepsilon) + W(\rho_\varepsilon))$$

on obtient :

$$\int_{\mathbb{R}^2} \frac{\partial \rho_\varepsilon}{\partial t} . (\frac{\partial W}{\partial \rho_\varepsilon}(\rho_\varepsilon)) \rho_\varepsilon + W(\rho_\varepsilon)) dx + \int_{\mathbb{R}^2} \nabla . (\rho_\varepsilon u_\varepsilon)(\rho_\varepsilon \frac{\partial W}{\partial \rho_\varepsilon}(\rho_\varepsilon) + W(\rho_\varepsilon)) dx = 0$$

Compte tenu des relations suivantes :

$$\rho_\varepsilon \frac{\partial W}{\partial \rho_\varepsilon}(\rho_\varepsilon) \frac{\partial \rho_\varepsilon}{\partial t} + W(\rho_\varepsilon) \frac{\partial \rho_\varepsilon}{\partial t} = \frac{\partial}{\partial t} (\rho_\varepsilon (W(\rho_\varepsilon))$$

$$\nabla . (\rho_\varepsilon u_\varepsilon) \rho_\varepsilon \frac{\partial W}{\partial \rho_\varepsilon}(\rho_\varepsilon) = \rho_\varepsilon^2 \nabla . u_\varepsilon \frac{\partial W}{\partial \rho_\varepsilon}(\rho_\varepsilon) + \rho_\varepsilon (u_\varepsilon . \nabla \rho_\varepsilon) \frac{\partial W}{\partial \rho_\varepsilon}(\rho_\varepsilon)$$

et $$\int_{\mathbb{R}^2} \nabla . (\rho_\varepsilon u_\varepsilon) W(\rho_\varepsilon) dx = -\int_{\mathbb{R}^2} ((\rho_\varepsilon u_\varepsilon) . \nabla \rho_\varepsilon) \frac{\partial W}{\partial \rho_\varepsilon}(\rho_\varepsilon) dx$$

On en déduit l'égalité suivante :

$$\frac{d}{dt} |\sqrt{\rho_\varepsilon W(\rho_\varepsilon)}|^2 + \int_{\mathbb{R}^2} p_\varepsilon \nabla . u_\varepsilon \, dx = 0 \tag{39}$$

d'où, en additionnant (37) et (39), on obtient l'égalité d'énergie :

$$\frac{d}{dt} \{ \frac{1}{2} |\sqrt{\rho_\varepsilon}\, u_\varepsilon|^2 + |\sqrt{\rho_\varepsilon W(\rho_\varepsilon)}|^2 \} + \nu |\nabla u_\varepsilon|^2 = 0 \tag{40}$$

Pour en déduire l'inégalité d'énergie, il reste à prouver deux lemmes. Le premier concerne un aspect important des problèmes de fluides compressibles et consiste à montrer que la densité est bornée supérieurement et inférieurement. Le deuxième porte sur une estimation de la quantité :

$$|\sqrt{\rho_\varepsilon\ W(\rho_\varepsilon)}|^2$$

qui nous permettra d'obtenir le terme :

$$\varepsilon|p_\varepsilon(t,.)|^2$$

qui apparaît dans (33).

<u>Lemme 2</u> : *Sous l'hypothèse H(A,T), il existe trois constantes positives* ρ_1, ρ_2, ε_1 *indépendantes de* ε *et de* T *telles que les inégalités suivantes aient lieu :*

$$\rho_1 \leqslant \rho_\varepsilon(t,x) \leqslant \rho_2 \qquad \forall (t,x) \in [0,T] \times \mathbb{R}^2 \tag{41}$$

<u>Preuve</u>

On utilise la formule $\rho_\varepsilon - \rho_0 = \varepsilon\, p_\varepsilon$ et la relation (30) rappelée dans le lemme (1), ce qui compte tenu de l'hypothèse H(A,T) donne :

$$\|\rho_\varepsilon(t,.)-\rho_0\|_{L^\infty(\mathbb{R}^2)} = \varepsilon\,\|p_\varepsilon(t,.)\|_{L^\infty(\mathbb{R}^2)} \leqslant b_2\sqrt{\varepsilon}\,\sqrt{\varepsilon}\,\|p_\varepsilon(t,.)\|_2 \leqslant b_2\sqrt{\varepsilon}\,\sqrt{A}\,.$$

On introduit ρ_1 fixe vérifiant $0 < \rho_1 < \rho_0$ et ε_1 défini par la formule :

$$\varepsilon_1 = \frac{(\rho_0 - \rho_1)^2}{b_2^2\, A}\,,$$

et on pose $\rho_2 = \rho_0 + b_2\,\varepsilon_1\sqrt{A}$. On a alors pour tout ε $(0 < \varepsilon < \varepsilon_1)$:

$$0 < \rho_1 \leqslant \rho_\varepsilon(t,x) \leqslant \rho_2\,.$$

On va maintenant montrer un lemme qui permettra d'estimer la quantité $\varepsilon|p_\varepsilon(t,.)|^2$ à partir de l'inégalité (40), en fonction du terme $|(\sqrt{\rho_\varepsilon\ W(\rho_\varepsilon)})(t,.)|^2$.

<u>Lemme 3</u> : *Il existe trois constantes positives indépendantes de* ε *et* T *telles que les inégalités suivantes aient lieu :*

$$C_4\,\varepsilon|p_\varepsilon(t,.)|^2 \leqslant \int_{\mathbb{R}^2} \rho_\varepsilon W(\rho_\varepsilon)dx < C_3\,\varepsilon|p_\varepsilon(t,.)|^2 \quad \forall t \in [0,T],\ \forall \varepsilon \leqslant \varepsilon_2 \tag{42}$$

Démonstration

Posons :

$$\Phi(\rho_\varepsilon) = \rho_\varepsilon\, W(\rho_\varepsilon) = \rho_\varepsilon \int_{\rho_o}^{\rho_\varepsilon} \frac{p_\varepsilon(s)}{s^2}\, ds$$

On a :

$$\Phi(\rho_o) = \Phi'(\rho_o) \qquad \text{et} \qquad \Phi''(\rho_\varepsilon) = \frac{1}{\varepsilon \rho_\varepsilon}$$

En faisant le développement de Taylor de Φ au voisinage de ρ_o, à l'ordre trois, on a :

$$\Phi(\rho_\varepsilon) = \frac{(\rho_\varepsilon - \rho_o)^2}{2\varepsilon\, \rho_o} - \frac{(\rho_\varepsilon - \rho_o)^3}{6\varepsilon\, \rho_\theta^2}$$

où

$$\rho_\theta = \rho_o + \theta(\rho_\varepsilon - \rho_o) \quad \text{avec} \quad \theta \in [0,1]$$

ce qui donne :

$$\Phi(\rho_\varepsilon) \geqslant \frac{(\rho_\varepsilon - \rho_o)^2}{2\varepsilon} \left[\frac{1}{\rho_o} - \frac{1}{3}\frac{(\rho_\varepsilon - \rho_o)}{\rho_\theta^2}\right].$$

On en déduit que :

$$J = \frac{1}{\rho_o} - \frac{1}{3}\ \frac{\rho_\varepsilon - \rho_o}{3\rho_\theta^2} \geqslant \alpha_1 > 0 \qquad \text{pour} \quad \rho_\varepsilon < \alpha_2$$

où, α_1 et α_2 sont deux constantes indépendantes de ε et t.

On choisit $\varepsilon \leqslant \varepsilon_2$ tel que $\rho_o + \sqrt{\varepsilon}\, b_2 \sqrt{A} < \alpha_2$.

D'où

$$\Phi(\rho_\varepsilon) \geqslant \alpha_1 \frac{(\rho_\varepsilon - \rho_o)^2}{2\varepsilon} \geqslant \frac{\alpha_1}{2}\, \varepsilon\, p_\varepsilon^2 .$$

Et donc, en intégrant sur $\mathbb{R}^2$, on obtient pour $\varepsilon \leqslant \varepsilon_1$

$$\int_{\mathbb{R}^2} \Phi(\rho_\varepsilon)dx = \int_{\mathbb{R}^2} \phi_\varepsilon\left[\int_{\rho_o}^{\rho_\varepsilon} \frac{p(s)}{s^2}\, ds\right]dx \leqslant \int_{\mathbb{R}^2} \rho_\varepsilon|\rho_\varepsilon - \rho_o|\ \left|\frac{p_\varepsilon(\rho_\varepsilon)}{\rho_o^2}\right|dx \leqslant \varepsilon\, \frac{\rho_2}{\rho_o^2}\, |p_\varepsilon(t,.)|^2 \qquad (43)$$

On pose $C_3 = \dfrac{\rho_2}{\rho_o^2}$ et on a :

$$\int_{\mathbb{R}^2} \Phi(\rho_\varepsilon)dx \geqslant \alpha_1\, \frac{\varepsilon}{2}\, |p_\varepsilon(t,.)|^2 \qquad (44)$$

En utilisant (43) et (44) et en posant $C_4 = \dfrac{\alpha_1}{2}$, on obtient les inégalités cherchée. Ceci termine la démonstration du lemme 3.

Enfin, on intègre (40) entre 0 et t et en appliquant les lemmes 2 et 3, on obtient la relation (33). Ceci complète la preuve du théorème 2.

I.3. Estimations sur les dérivées premières de u_ε et p_ε

On cherche à estimer les noms de $(u_\varepsilon, \sqrt{\varepsilon}\, p_\varepsilon)$ dans $L^\infty([0,T], H^1(\mathbb{R}^2))^2$.

C'est à ce niveau que nous introduisons la nouvelle inconnue $v_\varepsilon = \rho_\varepsilon . u_\varepsilon$ (qui n'est autre que le moment cinétique) et les équations :

$$\frac{\partial v_\varepsilon}{\partial t} + (v_\varepsilon.\nabla)u_\varepsilon + u_\varepsilon \nabla.v_\varepsilon - \nu\Delta u_\varepsilon = -\nabla p_\varepsilon \tag{45}$$

$$\varepsilon \frac{\partial p_\varepsilon}{\partial t} + \nabla.v_\varepsilon = 0 \tag{46}$$

En effet, en faisant les opérations classiques de dérivation et d'intégration sur le système initial, le terme ∇p_ε ne disparaît pas, or nous ne pouvons estimer que le terme $\sqrt{\varepsilon}\,\nabla p_\varepsilon$, ce qui nous donnerait des estimations sur $\sqrt{\varepsilon}\, u_\varepsilon$, donc dépendantes de ε.

On dérive en x les équations du système ci-dessus et on multiplie par $\partial_i v_\varepsilon$, où i varie de 1 à 2, on en déduit alors le théorème suivant :

Théorème 3 : *Il existe quatre constantes $\nu_2, a_7, a_8, \varepsilon_2$, indépendantes de ε et de T telles que l'estimation suivante ait lieu :*

$$\|u_\varepsilon(t,.)\|_1^2 + \varepsilon\|p_\varepsilon(t,.)\|_1^2 + \nu_2\int_0^t \|\nabla u_\varepsilon(\tau,.)\|_1^2 d\tau + \varepsilon\int_0^t |\nabla p_\varepsilon(\tau,.)|^2 d\tau < a_7\varepsilon^2\int_0^t |D^2 p_\varepsilon(\tau,.)|^2 d\tau$$
$$+ a_8 E_0 \qquad \forall t \in [0,T], \quad \forall \varepsilon \leqslant \varepsilon_2 \tag{47}$$

Démonstration

Elle se fait en trois étapes : dans la première on travaillera sur le système (45), (46) et on montrera des estimations sur $(v_\varepsilon, p_\varepsilon)$ en norme $H^1(\mathbb{R}^2)$.
Dans la seconde étape, on cherchera à estimer le terme :

$$\varepsilon \int_0^t |\nabla p_\varepsilon(\tau,.)|^2 \, d\tau$$

Et dans la troisième étape, on combinera les estimations trouvées avec l'inégalité d'énergie, on transformera v_ε en u_ε et on déduira (47).

Première Etape estimation de $(v_\varepsilon, p_\varepsilon)$ dans $H^1(\mathbb{R}^2)$:

Lemme 4 : *Sous l'hypothèse $H(A,T)$, il existe cinq constantes $a_1, a_2, a_3, \nu_1, \varepsilon_1$ indépendantes de ε et de T telles que l'inégalité suivante ait lieu :*

$$|\nabla v_\varepsilon(t,.)|^2+\varepsilon|\nabla p_\varepsilon(t,.)|^2+\nu_1'\int_0^t|D^2\ _\varepsilon(\tau,.)|^2 d\tau < |\nabla v_o|+a_3E_o+a_1\varepsilon^2\int_0^t|D^2p_\varepsilon(\tau,.)|^2$$
$$+\ a_2\ \varepsilon\int_0^t|\nabla p_\varepsilon(\tau,.)|^2\ d\tau \tag{48}$$

$\forall t\ \varepsilon\ [0,T]$, $\qquad\forall\varepsilon\leqslant\varepsilon_1$.

Démonstration

i) On applique l'opérateur ∂_i aux équations (45) et (46), ce qui donne :

$$\frac{\partial}{\partial t}\partial_i v_\varepsilon + (\partial_i v_\varepsilon\nabla)u_\varepsilon+((v_\varepsilon\nabla.)\partial_i u_\varepsilon))+(\partial_i u_\varepsilon)(\nabla.u_\varepsilon)+u_\varepsilon\nabla.(\partial_i v_\varepsilon)-\nu\Delta\partial_i u_\varepsilon = -\nabla\partial_i p_\varepsilon$$
$$\varepsilon\frac{\partial}{\partial t}\partial_i p_\varepsilon = -\nabla(\partial_i v_\varepsilon)$$

ii) On fait le produit scalaire dans $L^2(\mathbb{R}^2)$ des équations ci-dessus avec respectivement $\partial_i v_\varepsilon$, $\partial_i p_\varepsilon$:

$$\frac{1}{2}\frac{d}{dt}|\partial_i v_\varepsilon(t,.)|^2+\int_{\mathbb{R}^2}(\partial_i v_\varepsilon\nabla.)u_\varepsilon\partial_i v_\varepsilon dx+\int_{\mathbb{R}^2}(v_\varepsilon\nabla.)\partial_i u_\varepsilon\partial_i v_\varepsilon dx+\int_{\mathbb{R}^2}(\partial_i u_\varepsilon)\partial_i v_\varepsilon dx +$$
$$+\int_{\mathbb{R}^2}u_\varepsilon\nabla.(\partial_i v_\varepsilon)\partial_i v_\varepsilon dx-\nu\int_{\mathbb{R}^2}\Delta\partial_i u_\varepsilon\partial_i v_\varepsilon dx=-\int_{\mathbb{R}^2}\nabla\partial_i p_\varepsilon\partial_i v_\varepsilon dx \tag{49}$$

$$\frac{1}{2}\varepsilon\frac{d}{dt}|\partial_i p_\varepsilon(t,.)|^2 = -\int_{\mathbb{R}^2}\nabla.(\partial_i v_\varepsilon).\partial_i p_\varepsilon\ dx \tag{50}$$

On additionne (49) et (50), ce qui annule les seconds membres, et on remplace dans certains termes v_ε par $\rho_\varepsilon u_\varepsilon$, ce qui donne :

$$\frac{1}{2}\{|\nabla v_\varepsilon(t,.)|^2+\varepsilon|\nabla p_\varepsilon(t,.)|^2\}\ \nu\int_0^t|D^2u_\varepsilon(\tau,.)|^2d\tau\ \leqslant 2\int_0^t\int_{\mathbb{R}^2}|D^2u_\varepsilon||\nabla u_\varepsilon||\nabla\rho_\varepsilon|dx\ d\tau\ +$$
$$+\int_0^t\int_{\mathbb{R}^2}|D^2u_\varepsilon||D^2\rho_\varepsilon||u_\varepsilon|dx\,d\tau+4\int_0^t|u_\varepsilon|^2|\nabla\rho_\varepsilon||\nabla u_\varepsilon|dx\ d\tau +$$
$$+\ 6\int_0^t\int_{\mathbb{R}^2}|u_\varepsilon||\nabla\rho_\varepsilon||\nabla u_\varepsilon|^2\ \rho_\varepsilon\ dx\ d\tau\ +\ 2\int_0^t\int_{\mathbb{R}^2}|\rho_\varepsilon|^2|\nabla u_\varepsilon|^3\ dx\ d\tau\ +$$
$$+\ 2\int_0^t\int_{\mathbb{R}^2}|D^2u_\varepsilon||\nabla u_\varepsilon||\rho_\varepsilon|^2|u_\varepsilon|dx\ d\tau\ +\ 2\int_0\int_{\mathbb{R}^2}|D^2u_\varepsilon||\nabla\rho_\varepsilon||u_\varepsilon|^2|\rho_\varepsilon|dx\ d\tau\ +$$
$$+\int_0^t\int_{\mathbb{R}^2}|u_\varepsilon|^3|D^2\rho_\varepsilon|^2dx\ d\tau\ +\int_0^t\int_{\mathbb{R}^2}|u_\varepsilon|^2|D^2\rho_\varepsilon|\ |\nabla u_\varepsilon|\ |\rho_\varepsilon|dx\ d\tau\ +$$
$$+\ \frac{1}{2}\{|\nabla v_o|^2+\varepsilon|\nabla p_o|^2\}.$$

On obtient ainsi neuf termes différents ; on les estime en utilisant les inégalités (29), (30) du lemme 1 :

$$2\int_0^t\int_{\mathbb{R}^2}|D^2u_\varepsilon||\nabla u_\varepsilon||\nabla\rho_\varepsilon|dx\,d\tau\leqslant\frac{\nu\rho_1}{26}\int_0^t|D^2u_\varepsilon(\tau,.)|^2d\tau+\varepsilon^2b_1^2\ c(\nu,\rho_s)\int_0^t|\nabla p_\varepsilon||D^2p_\varepsilon|^2$$
$$|\nabla u_\varepsilon|^2d\tau \tag{51}$$

où $c(\nu,\rho_1)$ est une constante ne dépendant que de ν et de ρ_1.

Le second terme du second membre de (51) se majore en utilisant (32) et

(33). On a :

$$\varepsilon^2 \int_0^t |\nabla p_\varepsilon||D^2 p_\varepsilon|^2 |\nabla u_\varepsilon(\tau,.)|^2 \, d\tau <$$

$$\frac{C_1}{\nu_1} E_o . \sup_t \varepsilon^2 |\nabla p_\varepsilon(t,.)||D^2 p_\varepsilon(t,.)|^2$$

$$\leq (\sqrt{\varepsilon_1} \frac{C_1}{\nu_1} E_o)(\sqrt{\varepsilon} |\nabla p_\varepsilon(t,.)|)(\varepsilon |D^2 p_\varepsilon(t,.)|^2)$$

$$\leq \frac{C_1}{\nu_1} E_o \sqrt{\varepsilon_1} A \sqrt{A}$$

ainsi il vient :

$$2 \int_0^t \int_{\mathbb{R}^2} |D^2 u_\varepsilon||\nabla u_\varepsilon||\nabla \rho_\varepsilon| dx\, d\tau \leq \frac{\nu\rho_1}{26} \int_0^t |D^2 u_\varepsilon(\tau,.)|^2 d\tau + \varepsilon_1^2 b_1^2 c(\nu,\rho_1) A^2 \frac{C_1}{\nu_1} E_o$$

On travaille de la même manière pour les autres termes en séparant :

$$\int_0^t |D^2 u_\varepsilon(\tau,.)|^2 \, d\tau$$

des autres quantités.

De plus, en vertu du lemme 2, l'inégalité suivante a lieu :

$$\int_0^t \int_{\mathbb{R}^2} |D^2 u_\varepsilon(\tau,x)|^2 \rho_\varepsilon(\tau,x) dx\, d\tau \geq \rho_1 \int_0^t |D^2 u_\varepsilon(\tau,.)|^2 d\tau \ , \quad \forall \varepsilon \leq \varepsilon_1 \ ,$$

tout ceci nous permet d'obtenir la relation suivante :

$$\frac{1}{2}\frac{d}{dt}\{|\nabla v_\varepsilon(t,.)|^2 + \varepsilon|\nabla p_\varepsilon(t,.)|^2\} + \frac{\nu\rho_1}{2} \int_0^t |D^2 u_\varepsilon(\tau,.)|^2 d\tau \leq \frac{a_1}{2}\varepsilon^2 \int_0^t |D^2 p_\varepsilon(\tau,.)|^2 d\tau +$$
$$+ \frac{a_2}{2}\varepsilon \int_0^t |\nabla p_\varepsilon(\tau,.)|^2 d\tau + \frac{a_3}{2} E_o + \frac{|\nabla v_o|^2}{2}$$
$$\forall t \in [0,T], \quad \forall \varepsilon \leq \varepsilon_1.$$

où les constantes a_1, a_2, a_3 sont données par :

$$a_1 = 2(c(\nu,\rho_1) b_2^2 A + 1) \ , \quad a_2 = 2(4\alpha + \frac{\varepsilon_1}{2} b_2^6 A) \quad (\alpha \text{ constante positive arbitraire})$$

$$a_3 = 2\{2\varepsilon_1^2 b_1^8 c(\nu,\rho_1) \frac{A^2 C_1}{\nu_1} + 4c(\alpha)\varepsilon_1^4 b_1^8 b_2^8 \frac{A^6 C_1}{\nu_1} + 6\varepsilon_1 C(\nu,\rho_1) b_2^2 b_1^4 \rho_2^2 \frac{AC_1}{\nu_1} + 2\rho_2^2 b_1^2 \sqrt{A} \frac{C_1}{\nu_1} +$$
$$2C(\nu,\rho_1) b_2^2 \rho_2^4 \frac{AC_1}{\nu_1} + 2C(\nu,\rho_1) b_2^2 b_1^4 \rho_2^2 A^2 \frac{C_1}{\nu_1} + \varepsilon_1^2 b_2^4 \rho_2^2 \frac{A^2 C_1}{\nu_1} + 1\} \ .$$

Ceci donne l'estimation cherchée en posant $\nu' = \nu\rho_1$.

<u>Deuxième étape estimation de la norme de</u> $\sqrt{\varepsilon}\, \nabla p_\varepsilon$ <u>dans</u> $L^2([0,T] \, ; L^2(\mathbb{R}^2))$

On se propose d'estimer le second membre de l'équation (9) et pour cela on prouve le :

<u>*Lemme 5*</u> *: Sous l'hypothèse* $H(A,T)$*, on peut trouver trois constantes* a_4*,* a_5*,* a_6 *indépendantes de* ε *et de* T *telles que l'on ait pour* $t \in [0,T]$ *et* $\varepsilon \leq \varepsilon_1$*,*

l'inégalité :

$$\varepsilon\int_o^t |\nabla p_\varepsilon(\tau,.)|^2 d\tau \leqslant a_4\left[\int_{\mathbb{R}^2} p_\varepsilon.\nabla u_\varepsilon dx\right]_o^t + \varepsilon\, a_3\int_o^t \; D^3 u\;(\;,.)\;{}^2 d\; + a_6 E_o \qquad (52)$$

Démonstration

Comme ρ_ε est strictement positif, on peut multiplier scalairement l'équation (9) par $-\varepsilon\,\nabla p_\varepsilon/\rho_\varepsilon$ ce qui donne, après intégration sur $|0,T|\times\mathbb{R}^2$:

$$\varepsilon\int_o^t\int_{\mathbb{R}^2}\frac{|\nabla p_\varepsilon|}{\rho_\varepsilon}dx\,d\tau = -\varepsilon\int_o^t\int_{\mathbb{R}^2}\frac{\partial u_\varepsilon}{\partial t}\nabla p_\varepsilon\,dx\,d\tau - \varepsilon\int_o^t\int_{\mathbb{R}^2}(u_\varepsilon.\nabla)u_\varepsilon\nabla\;\;{}_\varepsilon dx\,d\tau + \nu\varepsilon\int_o^t\int_{\mathbb{R}^2}\Delta u_\varepsilon\frac{\nabla P_\varepsilon}{\rho_\varepsilon}dx\,d\tau$$

On estime chaque terme séparément ; pour le premier, on a pour tout $\varepsilon<\varepsilon_1$:

$$\int_o^t\int_{\mathbb{R}^2}\frac{|\nabla p_\varepsilon|^2}{\rho_\varepsilon}dx\,d\tau > \frac{1}{\rho_2}\int_o^t |\nabla p_\varepsilon(\tau,.)|^2\,d\tau$$

pour le second terme, on a :

$$\varepsilon\int_o^t\int_{\mathbb{R}^2}\frac{\partial u_\varepsilon}{\partial t}.\nabla p_\varepsilon\,dx\,d\tau = -\int_o^t\int_{\mathbb{R}^2} p_\varepsilon\frac{\partial}{\partial t}\nabla.u_\varepsilon\,dx\,d\tau$$

De plus, en vertu de (10), il vient :

$$\varepsilon\frac{\partial}{\partial t}(p_\varepsilon.\nabla.u_\varepsilon) = \varepsilon\, p_\varepsilon(\frac{\partial}{\partial t}(\nabla.u_\varepsilon)) - (u_\varepsilon\nabla\rho_\varepsilon)\nabla.u_\varepsilon - \rho_\varepsilon(\nabla.u_\varepsilon)^2$$

d'où

$$\int_o^t\int_{\mathbb{R}^2}\varepsilon\frac{\partial}{\partial t}u_\varepsilon\nabla p_\varepsilon\,dx\,d\tau = \int_o^t\frac{\partial}{\partial t}\left[\int_{\mathbb{R}^2}\varepsilon\,p_\varepsilon\,\nabla.u_\varepsilon\,dx\right]d\tau + \int_o^t\int_{\mathbb{R}^2}\nabla.(\rho_\varepsilon u_\varepsilon)\nabla.u_\varepsilon\,dx\,d\tau \quad .$$

or on a :

$$\int_o^t\int_{\mathbb{R}^2}\nabla.(\rho_\varepsilon u_\varepsilon)\nabla.u_\varepsilon\,dx\,d\tau \leqslant \frac{\varepsilon}{6\rho_2}\int_o^t|\nabla p_\varepsilon(\tau,.)|^2 d\tau + (\rho_2 + C(\rho_2^{-1})b_2^2 A)\frac{C_1}{\nu_1}E_o$$

En raisonnant de la même manière pour les autres termes, on en déduit l'inégalité suivante (valable pour $\varepsilon<\varepsilon_1$ et $t\in[0,T]$) :

$$\frac{\varepsilon}{2\rho_2}\int_o^t|\nabla p_\varepsilon(\tau,.)|^2 d\tau < \left[\int_{\mathbb{R}^2}p_\varepsilon\,\nabla.u_\varepsilon\,dx\right]_o^t + \frac{\nu^2}{\rho_1^2}\varepsilon\,c(\rho_2^{-1})\int_o^t|D^2u_\varepsilon(\tau,.)|^2 d\tau + $$
$$+ (2b_2^2\;AC(\rho_2^{-1})+\rho_2)\frac{C_1}{\nu_1}E_o\;.$$

L'inégalité (52) se déduit immédiatement en posant :

$$a_4 = 2\rho_2\;,\qquad a_5 = \frac{2\nu^2\varepsilon}{\rho_1^2}\rho_2\;C(\rho_2^{-1})\;,\qquad a_6 = (2b_2^2\;AC(\rho_2^{-1}) + \rho_2)\frac{C_1}{\nu_1}\rho_2$$

Troisième étape - Conclusion

On a maintenant les matériaux pour obtenir l'estimation énoncée dans le théorème 3 :

i) On exprime le terme $|\nabla v_\varepsilon|^2$ qui apparaît dans l'inégalité (51) en fonction de u_ε, ρ_ε :

$$|\nabla v_\varepsilon(t,.)|^2 = \sum_{i,j=1}^{2} \int_{\mathbb{R}^2} \{(u_{\varepsilon,j}\ \partial_i\rho_\varepsilon)^2 + 2(u_{\varepsilon,j}\partial_i\rho_\varepsilon)(\partial_i u_{\varepsilon,j}\rho_\varepsilon) + \rho_\varepsilon\partial_i u_{\varepsilon,j}\}(t,x)dx$$

et on utilise les inégalités suivantes :

$$\sum_{i,j=1}^{2} \int_{\mathbb{R}^2} (u_{\varepsilon,j}\partial_i\rho_\varepsilon)^2(t,x)dx \leq \varepsilon^2\, \|u_\varepsilon(t,.)\|^2_{L^\infty(\mathbb{R}^2)}\, |\nabla p_\varepsilon(t,.)|^2 \leq b_2\ A|\nabla p_\varepsilon(t,.)|^2$$

$$-2\sum_{i,j=1}^{2} \int_{\mathbb{R}^2} (u_{\varepsilon,j}\partial_i\rho_\varepsilon)(\rho_\varepsilon\partial_i u_{\varepsilon,j})dx \leq \frac{\varepsilon}{4}\, |\nabla p_\varepsilon(t,.)|^2 + 8\varepsilon\ b_2^2\ \rho_2^2\ A\ |\nabla u_\varepsilon(t,.)|^2$$

ceci $\forall\varepsilon \leq \varepsilon_1$, et

$$p_1^2|\nabla u_\varepsilon(t,.)|^2 \leq \sum_{i,j=1}^{2} \int_{\mathbb{R}^2} \rho_\varepsilon(\partial_i u_{\varepsilon,j})^2\, dx \leq \rho_2^2|\nabla u_\varepsilon(t,.)|^2 .$$

ii) On estime, dans (52), le terme $\int_{\mathbb{R}^2} p_\varepsilon\ \nabla.u_\varepsilon(t,x)dx$; on a :

$$\varepsilon\ a_4 \int_{\mathbb{R}^2} p_\varepsilon\ \ \nabla.u_\varepsilon(t,x)dx \leq \frac{\varepsilon}{4}|p_\varepsilon(t,.)|^2 + \varepsilon\ a_4^2|\nabla u_\varepsilon(t,.)^2|$$

iii) On additionne l'inégalité d'énergie (33) avec les inégalités des lemmes 4 et 5, ce qui donne :

$$|u_\varepsilon(t,.)|^2+(\rho_1-8\varepsilon b_2^2\rho_2^2A-4\varepsilon a_4^2)|\nabla u_\varepsilon(t,.)|^2+ \frac{\varepsilon}{2}|p_\varepsilon(t,.)|^2+\varepsilon|\nabla p_\varepsilon(t,.)|^2+\nu \int_0^t|\nabla u_\varepsilon(\tau,.)|^2 d\tau$$
$$+ \nu_1' \int_0^t|D^2u_\varepsilon(\tau,.)|^2d\tau+\varepsilon\int_0^t|\nabla p_\varepsilon(\tau,.)|^2d\tau \leq a_1\varepsilon^2\int_0^t|D^2p_\varepsilon(\tau,.)|^2 + a_8\ E_o$$

où

$$a_8 = \mathrm{Max}\{a_3 + a_6\ ,\ \frac{1}{4}+\varepsilon\, b_2^2\, A\ ,\ \frac{1}{4}\ ,\ 8\varepsilon\ b_2^2\ \rho_2^2\ A + 1\ ,\ \ 4\varepsilon a_1^2\}$$

On choisit ε_2' tel que l'on ait pour $\varepsilon \leq \varepsilon_2'$:

$$\rho_1 - 8\varepsilon\ b_2^2\ \rho_2^2\ A - 4\varepsilon\ a_4^2 \geq \gamma > 0\ ,$$

et on pose $\varepsilon_2 = \min(\varepsilon_1,\varepsilon_2')$. On en déduit les constantes a_7, ν_2' de l'inégalité cherchée.

I.4. Estimations sur les dérivées secondes de $(u_\varepsilon,p_\varepsilon)$

On réitère le raisonnement précédent en l'appliquant aux dérivées secondes des équations (45) et (46) ; on montre donc le :

Théorème 4 : *Sous l'hypothèse $H(A,T)$, il existe quatre constantes ν_3, a_9, a_{10}, ε_3, indépendantes de ε et de T, telles que pour tout $t \in [0,T]$ et tout $\varepsilon \in]0,\varepsilon_3]$ on ait :*

$$\|u_\varepsilon(t,.)\|_2^2+\varepsilon\|p_\varepsilon,.)\|_2^2+\nu_3\int_0^t\|\nabla u_\varepsilon(\tau,.)\|_2^2\,d\tau+\varepsilon\int_0^t\|\nabla p_\varepsilon(\tau,.)\|_1^2 d\tau$$
$$\leqslant a_9\,\varepsilon^{3/2}\int_0^t|D^2p_\varepsilon(\tau,.)|^2d\tau+a_{10}\,E_o \tag{53}$$

Première étape

On travaille sur les équations (45) et (46) et on montre le lemme suivant :

Lemme 6 : Il existe quatre constantes positives, indépendantes de ε *et* T, *telles que* $(v_\varepsilon,p_\varepsilon,u_\varepsilon)$ *vérifient pour* $t\in[0,T]$ *et* $\varepsilon\in\,]0,\varepsilon_3']$, *l'inégalité suivante :*

$$|D^2v_\varepsilon(t,.)|^2+\varepsilon|D^2p_\varepsilon(t,.)|^2+\nu_3'\int_0^t|D^3u_\varepsilon(\tau,.)|^2\leqslant a_{11}\varepsilon^2\int_0^t\|\nabla p_\varepsilon(\tau,.)\|_1^2 d\tau$$
$$+a_{12}\,\varepsilon\,\sqrt{\varepsilon}\int_0^t|D^3p_\varepsilon(\tau,.)|^2d\tau+a_{13}\,E_o+|D^2\,\sqrt{o}|^2$$

Démonstration

Elle se fait comme celle du lemme 4. Par les mêmes raisonnements on obtient 19 termes à estimer. Ce que l'on fait à l'aide des inégalités du lemme 1, en veillant à isoler le terme $\int_0|D^3u_\varepsilon(\tau,.)|^2d\tau$. On obtient ainsi un coefficient en $\varepsilon\sqrt{\varepsilon}$ pour le terme $\int_0^t|D^3p_\varepsilon(\tau,.)|^2d\tau$.

Deuxième étape

On cherche une estimation sur $\int_0^t|D^2p_\varepsilon(\tau,.)|^2d\tau$, et on montre le lemme suivant :

Lemme 7 : Il existe trois constantes a_{14}, a_{15}, ε_3^4, *indépendantes de* ε *et de* T *telles que pour tout* $t\in[0,T]$ *et* $\varepsilon\in\,]0,\varepsilon_3'']$ *on ait :*

$$\varepsilon\int_0^t|D^2p_\varepsilon(\tau,.)|^2d\tau+\varepsilon\Big[\int_{\mathbb{R}^2}(\nabla u_\varepsilon,p_\varepsilon)(x,.)dx\Big]_0^t\leqslant a_{14}\int_0^t|D^3u_\varepsilon(\tau,.)|^2d\tau+a_{15}E_o \tag{55}$$

Démonstration

On procède comme dans le lemme 5, en dérivant en x, l'équation (9).

Troisième étape

On additionne les équations (55) et (54) et l'équation (47), multipliée par une constante positive ∂, que l'on choisira ultérieurement.

Par ailleurs, comme dans le paragraphe (II.3), on montre le :

Lemme 8 : *Sous l'hypothèse* $H(A,T)$, *il existe deux constantes indépendantes de* ε *et* T, *telles que pour tout* $t \in [0,T]$ *et* $\varepsilon \in]0,\varepsilon_1]$, *on ait* :

$$- \frac{\varepsilon}{4} |D^2 p_\varepsilon(t,.)|^2 + (\rho_2 - d_2\sqrt{\varepsilon})|D^2 u_\varepsilon(t,.)|^2 \leqslant |D^2 v_\varepsilon(t,.)|^2$$

$$\leqslant (\frac{1}{3} + \varepsilon b_2 A)\varepsilon |D^2 p_\varepsilon(t,.)|^2 + d_1 \sqrt{\varepsilon}\, |D^2 u_\varepsilon(t,.)|^2 \qquad (56)$$

Démonstration

On procède comme pour le théorème 3 en utilisant les inégalités du lemme 1 et l'hypothèse $H(A,t)$.

En tenant compte du lemme 7, on choisit γ assez grand et ε assez petit. On obtient l'inégalité (53), avec les constantes ν_3, a_9, a_{10}, ε_3 , indépendantes de ε et de T.

I.5. Estimations des dérivées troisièmes de $(u_\varepsilon, p_\varepsilon)$

On va de nouveau considérer le système initial (9), (10) et y appliquer les techniques de Nishida, Matsumura [15]. Ceci va nous permettre d'obtenir des estimations qui ne dépendent pas du terme $\int_0^t |D^2 p_\varepsilon(\tau,.)|^2 d\tau$. Mais nous ne pourrons les obtenir que sur les quantités $\varepsilon|D^3 u_\varepsilon(t,.)|^2$ et $\varepsilon^2 |D^3 p_\varepsilon(t,.)|^2$. Ce qui donne le :

Théorème 5 : *Sous l'hypothèse* $H,A,T)$, *il existe quatre constantes* C_2, ν_4', ε_4, α_o , *indépendantes de* ε *et de* T, *telles que pour* $t \in [0,T]$ *et* $\varepsilon \in]0,\varepsilon_4]$, *on ait* :

$$\|u_\varepsilon(t,.)\|_2^2 + \varepsilon \|p_\varepsilon(t,.)\|_2^2 + \varepsilon^2 |D^3 p_\varepsilon(t,.)|^2 + \alpha_o \varepsilon \int_0^t \|\nabla p_\varepsilon(\tau,.)\|_2^2\, d\tau$$

$$+ \nu_4' \int_0^t \|\nabla u_\varepsilon(\tau,.)\|_2^2\, d\tau \leqslant C_2\, E_o \qquad (57)$$

Démonstration

i) On applique l'opérateur $\partial_j\, \partial_k\, \partial_i$ à l'équation (10) et on fait le produit scalaire dans $L^2(\mathbb{R}^2)$ avec $\nu\, \varepsilon\, \partial_j\, \partial_k\, \partial_i\, p_\varepsilon$.

ii) On applique l'opérateur $\partial_k\, \partial_i$ à l'équation (9), qui a été, auparavant divisé par ρ_ε ; et on fait le produit scalaire avec $\nu\varepsilon\, \rho_\varepsilon\, \nabla\partial_k\, \partial_i\, p_\varepsilon$.

iii) On applique l'opérateur $\partial_j \partial_k \partial_i$ à l'équation (10), on multiplie dans $L^2(\mathbb{R}^2)$ avec $\rho_\varepsilon^2 \partial_k \partial_i u_{\varepsilon,j}$ et on somme pour ∂ variant de 1 à 2.

iv) On multiplie par le terme $2\varepsilon\rho_\varepsilon(\partial_k \partial_i u_\varepsilon).(\nabla\partial_k \partial_i p_\varepsilon)$ et on intègre sur $\mathbb{R}^2$.

v) On somme les quatre équations obtenues, les dérivées d'ordre 4 en u_ε disparaissent. On obtient :

$$(\varepsilon^2 \frac{\nu}{2} - \varepsilon^2\alpha)|D^2 p_\varepsilon(t,.)|^2 - \rho_2(\alpha)|D^2 u_\varepsilon(t,.)|^2 + \varepsilon(\frac{\rho_1}{2} - \frac{7}{2}\sqrt{\varepsilon}\, b_2\sqrt{A}\,\nu)\int_0^t |D^3 p_\varepsilon(\tau,.)|^2 d\tau$$
$$- \rho_2^3 \int_0^t |D^3 u_\varepsilon(\tau,.)|^2 d\tau \leq \alpha_1\sqrt{\varepsilon}\int_0^t \|\nabla u_\varepsilon(\tau,.)\|_1^2\, d\tau + d_2^1\,\varepsilon\sqrt{\varepsilon}\int_0^t \|\nabla p_\varepsilon(\tau,.)\|_1^2 d\tau + \alpha_3^1 E_o \quad (58)$$

α et donc $C(\alpha)$ sont choisis de manière à vérifier $\frac{\nu}{2} - \alpha > 0$ et d_1', d_2', d_3' sont des constantes indépendantes de ε de T.

On somme alors l'inégalité (58) avec γ fois l'inégalité du Théorème 4. On choisit γ assez grand de telle sorte que l'on ait $\gamma\nu_3 - \rho_2^3 \geqslant \nu_4' > 0$ et $\gamma - C(\alpha)\rho_2^2 > 1 > 0$. On choisit alors ε assez petit et on déduit l'estimation (57).

On cherche alors les estimations sur les dérivées de u_ε :

<u>*Théorème 6*</u> *: Sous l'hypothèse $H(A,T)$ il existe deux constantes ν_4'', C_2' indépendantes de ε et de T telles que pour $t \in [0,T]$ et $\varepsilon \in]0,\varepsilon_4]$, on ait :*

$$\varepsilon|D^3 u_\varepsilon(t,.)|^2 + \nu_4''\,\varepsilon\int_0^t |D^4 u_\varepsilon(\tau,.)|^2 d\tau \leq C_2' E_o \quad (59)$$

<u>Démonstration</u>

On applique l'opérateur $\partial_k \partial_i$ à l'équation (9) et on multiplie scalairement dans $L^2(\mathbb{R}^2)$ l'équation obtenue par $-\varepsilon\,\Delta\,\partial_k \partial_i u_\varepsilon$. On raisonne, comme précédemment, et on déduit (59).

I.6. <u>Obtention de l'estimation indépendante de ε et l'existence globale en temps</u>

Les estimations à priori obtenues ci-dessus suffisent à prouver l'existence globale d'une solution bornée dans les espaces convenables, indépendamment de ε, c'est l'objet du :

<u>*Théorème 7*</u> : *Pour toute donnée initiale* $(u_o, p_o) \in (H^3(\mathbb{R}^2))^2 \times H^3(\mathbb{R}^2)$ *on désigne par* E_o *l'expression :*

$$E_o = \|u_o\|_2^2 + \varepsilon\, |D^3 u_o|^2 + \varepsilon\, \|p_o\|_2^2 + \varepsilon^2 |D^3 p_\varepsilon(0,.)|^2 ,$$

et pour toute fonction $(u_\varepsilon, p_\varepsilon)$ *définie sur* $\mathbb{R}_+$ *à valeur dans* $(H^3(\mathbb{R}^2))^2 \times H^3(\mathbb{R}^2)$ *on désigne par* $E_\varepsilon(t)$ *l'expression :*

$$E_\varepsilon(t) = \|u_\varepsilon(t,.)\|_2^2 + \varepsilon |D^3 u_\varepsilon(t,.)|^2 + \varepsilon\, \|p_\varepsilon(t,.)\|_2^2 + \varepsilon^2 |D^3\, p_\varepsilon(t,.)|^2 .$$

Alors si la fonction u_o *est à divergence nulle et si* E_o *est assez petit (par rapport à* ν*), le problème (9), (10), (11), (12) admet, pour tout* ε *assez petit une unique solution* $(u_\varepsilon(t,.), p_\varepsilon(t,.))$ *définie pour tout* t *et qui vérifie de plus une majoration à priori indépendantes de* ε *de la forme suivante :*

$$E_\varepsilon(t) + \nu_4 \int_0^t \|\nabla u_\varepsilon(\tau,.)\|_2^2 d\tau + \nu_4\, \varepsilon \int_0^t |D^4 u_\varepsilon(\tau,.)|^2 d\tau + \varepsilon \int_0^t \|\nabla p_\varepsilon(\tau,.)\|_2^2 d\tau \leq C\, E_o \qquad (60)$$

où C *et* ν_4 *sont des constantes indépendantes de* ε *et de* T.

<u>Démonstration</u>

Il existe A et T tels que le problème admette une solution appartenant à H(A,T). En additionnant les équations (57) et (59) on en déduit que cette solution vérifie sur l'intervalle $[0,T]$

$$E_\varepsilon(t) + \nu_4 \int_0^t \|\nabla u_\varepsilon(\tau,.)\|_2^2 d\tau + \nu_4\, \varepsilon \int_0^t |D^4 u_\varepsilon(\tau,.)|^2 d\tau + \varepsilon \int_0^\tau \|\nabla p_\varepsilon(\tau,.)\|_2^2 d\tau \leq C(A)\, E_o \quad (61)$$

où C(A) est une constante convenable, indépendante de ε et de T. Maintenant si E_o est choisi de manière à vérifier :

$$E_o \leq \min\left(\frac{A}{4}, \frac{A}{4C(A)}\right)$$

On montre que la solution est définie sur l'intervalle $[0,2T]$ et vérifie sur cet intervalle la majoration (61). Ainsi, comme dans Nishida-Matsumura, on obtient la preuve du théorème 7.

II. <u>Passage à la limite pour ε tendant vers zéro</u>

De l'inégalité (60) on en déduit que la suite $\{(u_\varepsilon, \sqrt{\varepsilon}\, p_\varepsilon\}_{\varepsilon \leq \varepsilon_5}$ est bornée dans :

$$(L^{\infty}([0,T] ; H^2(\mathbb{R}^2))^3 \cap (L^2([0,T] ; H^3(\mathbb{R}^2))^3 .$$

Et donc converge faiblement dans ces espaces, pour obtenir la convergence forte sur u_ε, nécessaire pour le passage à la limite dans le terme non linéaire $(\rho_\varepsilon u_\varepsilon.\nabla)u_\varepsilon$, il faut montrer une estimation sur la dérivée en temps de u_ε.

C'est l'objet du :

<u>Théorème 8</u> : *Sous les hypothèses du théorème 7, on peut trouver trois constantes K, ν_5, ε_5 indépendantes de ε telles que l'on ait, pour tout $t \in [0,T]$ et $\varepsilon \in]0,\varepsilon_5]$:*

$$\varepsilon|p'_\varepsilon(t,.)|^2 + |u'_\varepsilon(t,.)|^2 + \nu_5 \int_0^t |\nabla u'_\varepsilon(\tau,.)|^2 d\tau \leq K(T) \tag{62}$$

<u>Démonstration</u>

On applique l'opérateur $\frac{\partial}{\partial t}$ aux équations en $(v_\varepsilon, p_\varepsilon)$. On multiplie les équations obtenues par respectivement v'_ε et p'_ε

$$v'_\varepsilon = \frac{\partial v_\varepsilon}{\partial t} \quad \text{et} \quad p'_\varepsilon = \frac{\partial p_\varepsilon}{\partial t}$$, et on remarquera que

$v'_\varepsilon = -u_\varepsilon \nabla.v_\varepsilon - (v_\varepsilon.\nabla)u_\varepsilon + \nu\Delta u_\varepsilon - \nabla p_\varepsilon$ et $p'_\varepsilon = -\nabla.v_\varepsilon$ donc, en vertu des résultats du paragraphe I, on en déduit que v'_ε et p'_ε sont uniformément bornés dans $L^{\infty}([0,T] ; L^2(\mathbb{R}^2))$.

Par ailleurs, on a le lemme suivant :

<u>Lemme 9</u> : *Sous les hypothèses du théorème 8, $\rho'_\varepsilon = \frac{\partial \rho_\varepsilon}{\partial t}$ est borné dans $H^1(\mathbb{R}^2)$ par une constante K_o indépendante de t et de ε assez petit.*

<u>Démonstration</u>

De l'équation :

$$\frac{\partial \rho_\varepsilon}{\partial t} = -\nabla.(\rho_\varepsilon u_\varepsilon)$$

On en déduit la relation :

$$|\rho'_\varepsilon(t,.)| \leq (b_1^2\, C(A)\, E_o + \rho_2^2 + 2\rho_2) \sqrt{C(A)\, E_o}$$

Tandis que de l'équation :

$$\frac{\partial}{\partial t} \nabla\rho_\varepsilon + \nabla(\nabla.(\rho_\varepsilon u_\varepsilon)) = 0$$

on déduit la relation :

$$|\nabla\rho_\varepsilon(t,.)|^2 \leqslant K_1 \sqrt{C(A)\, E_o}$$

Comme les constantes b_1, $C(A)$, E_o, ρ_2 et k_1 sont indépendantes de ε et de t, cela démontre le lemme 9.

En vertu de la majoration obtenue sur ρ'_ε on en déduit que pour tout $\varepsilon \leqslant \varepsilon_5$

$$|v'_\varepsilon(t,.)|^2 \leqslant \int_o^t K_1(1 + \|\nabla u_\varepsilon(\tau,.)\|_2)\, |v'_\varepsilon(\tau,.)|^2 d\tau + K_2 t + K_3 \qquad (63)$$

On applique alors le lemme de Cronwall et on déduit que :

$$|v'_\varepsilon(t,.)|^2 \leqslant K_3(t) \exp K_1 T \quad ;$$

par le même raisonnement, on obtient que :

$$\int_o^t |\nabla u'_\varepsilon(\tau,.)|^2\, d\tau < K'(T) \quad ;$$

comme on a :

$$v'_\varepsilon = \rho'_\varepsilon\, u_\varepsilon + \rho_\varepsilon . u'_\varepsilon$$

des inégalités précédentes, on en déduit que, pour $\varepsilon \leqslant \varepsilon_6$, on a :

$$|u'_\varepsilon(t,.)|^2 + \varepsilon|p'_\varepsilon(t,.)|^2 + \nu_o - \int_o^t |\nabla u'_\varepsilon(\tau,.)|^2\, d\tau \leqslant K(T)$$

où ν_5 est une constante indépendante de ε . Ainsi la suite $\{u'\}_{\varepsilon\leqslant\varepsilon_6}$ est bornée dans $(L^\infty([0,T] \,;\, L^2(\mathbb{R}^2)))^2$ et $(L^2([0,T] \,;\, H'(\mathbb{R}^2))^2$.

Des théorème 7 et 8, on en déduit que la suite $\{u_\varepsilon\}_{\varepsilon\leqslant\varepsilon_6}$ et pour tout $T > 0$, bornée dans $(H'([0,T] \times \mathbb{R}^2))^2$ et que la suite $(\sqrt{\varepsilon}\, p_\varepsilon)_{\varepsilon\leqslant\varepsilon_6}$ est bornée dans $(H'[0,T] \times \mathbb{R}^2)$.

Ce qui, d'après le théorème de Reelich-Kondvatchov, donne une convergence forte de u_ε vers u_* dans $L^2(K)$ où K est un compact quelconque de $\mathbb{R}^+ \times \mathbb{R}^2$ d'où le :

<u>*Théorème 9*</u> : *On désigne par* $(u_\varepsilon, p_\varepsilon)_{\varepsilon\leqslant\varepsilon_6}$ *la solution du problème (9), (10), (11), (12). On suppose que les hypothèses du théorème 8 sont satisfaites, alors*

lorsque ε *tend vers zéro* $(u_\varepsilon, p_\varepsilon)$ *converge vers* (u,p) *solution des équations (5), (6) et (7) dans le sens suivant : pour tout* $T > 0$, u_ε *converge fortement vers* u *dans* $(L^2[0,T] \; ; \; L^2(\mathbb{R}^2))^2$ *et faiblement dans* $(L^\infty([0,T];H^2(\mathbb{R}^2)))^2 \cap$ $L^2([0,T] \; ; \; H^3(\mathbb{R}^2))^2$; *tandis que* ∇p_ε *converge faiblement vers* ∇p *dans* $L^\infty([0,T] \; ; \; L^2(\mathbb{R}^2))$ *et* $L^2[0,T] \; ; \; H'(\mathbb{R}^2))$.

Démonstration

Des convergences faibles et fortes de u_ε vers u_* découlent des estimations précédentes et des lemmes de compacité [12]. Pour montrer que u_* est solution des équations (5), (6) et (7), on procède comme dans [17] , pour toute fonction $\mathcal{C} \in \mathcal{D}([0,T] \times \mathbb{R}^2)$ à divergence nulle, on a :

$$< \frac{\partial v_\varepsilon}{\partial t} , \mathcal{C} > + <(v_\varepsilon . \nabla) u_\varepsilon , \mathcal{C} > + <u_\varepsilon \; \nabla . v_\varepsilon , \mathcal{C} > - \nu <\Delta u_\varepsilon , \mathcal{C} > = 0 \qquad (64)$$

On passe alors à la limite sur ε dans (64). De plus, pour tout $\mathcal{C} \in \mathcal{D}\,[0,T] \times \mathbb{R}^2)$ on a $\sqrt{\varepsilon} <\sqrt{\varepsilon} , \frac{\partial p_\varepsilon}{\partial t} , \mathcal{C} > + <\nabla . v_\varepsilon , \mathcal{C} > = 0$ soit à la limite $<\nabla v_* , \mathcal{C} > = 0$.

De la relation $\rho_\varepsilon - \rho_o = \sqrt{\varepsilon} . \sqrt{\varepsilon} \; p_\varepsilon$, on déduit que ρ_ε tend vers ρ_o uniformément sur tout compact de $\mathbb{R}_t^+ \times \mathbb{R}^2$. Ainsi u_* est la solution des équations de Navier-Stokes incompressibles :

$$\frac{\partial u_*}{\partial t} + (u_* . \nabla) u_* - \frac{\nu}{\rho_o} \Delta u_* = -\nabla p_* , \quad \nabla . u_* = 0 \; , \quad u_*(x,o) = v_o(x) \; .$$

Enfin de la relation :

$$\nabla p_\varepsilon = - \rho_\varepsilon \left(\frac{\partial u_\varepsilon}{\partial t} \right) + (u_\varepsilon . \nabla) u_\varepsilon) + \nu \; \Delta u_\varepsilon$$

on déduit que ∇p_ε converge vers ∇p_* , ce qui termine la démonstration du théorème 9.

Remarque : Nous avons montré la convergence globale en temps de la solution des équations d'un fluide compressible visqueux vers celle d'un fluide incompressible visqueux. Le cas étudié est en fait une approximation assez grossière de la notion physique de compressibilité. Contrairement à Klainerman et Majda qui ont étudié de façon précise les propriétés physiques de la compressibilité (associée au nombre de Mach) ce qui leur donne un système plus compliqué, sur lequel ils obtiennent un résultat local en temps.

Références

(1) EBIN D. : The motion of sslighty compressible fluide vewed as a motion with strong constraining force. Annals of mathematics 105 (1977) 141.200.

(2) FRIEDMAN A. : Partial Differential equation of parabolic type - Prince-Hall, Inc. (1964).

(3) ITAYA N. : On the Cauchy problem for the system of fondamental equations describing the movement of compressible viscous fluid - Kodaï Math. Sem. Rep. 23 (1971) 60-120.

(4) ITAYA N. : On the initial value problem of the motion of compressible viscous fluid, especially on the problem of uniqueness - J. Math. Kyoto Univ. 16 (1976) 413-427.

(5) ITAYA N. : A survey on the generalized Burger's equation with a pressure model term - J. Math. Kyoto Univ., 16 (1976) 1-18.

(6) ITAYA N. : A survey of two model equations for compressible viscous fluid (to appear).

(7) KANEL Ya.I. : On a model system of equations for one dimensional gas motion - Diff. eq. (in Russian) 8 (1968) 721-734.

(8) KATO T. : Linear evolution equations of hyperbolic type - J. Fac. Sci. Univ. Tokyo, 17 (1970) 241-258.

(9) KAZHYKOV A.V. : Sur la solubilité globale du problème monodimensionnel aux valeurs initiales limitées pour les équations du gaz visqueux et calorifique - CR Accord Sci. Paris 284 (1977) Ser. A 317.

(10) KAZHYKOV A.V. and SELUKNIN V.V. : Unique global solution in times of initial boundary value problems for one dimensional equations of a viscous gas P. M. M. Vol. 41, n° 2 (1977) Novosibirsk .

(11) LADYZHENSKAYA O.A.: The mathematical theorie of viscous incompressible flow. Gordon and Breach (2 edition) (1969).

(12) LIONS J.L. : Quelques méthodes de résolution des problèmes aux limites non linéaires - Dunod (1969).

(13) LIONS J.L. et MAGENES E. : Problèmes aux limites non homogènes - Volume 1 - Dunod (1968) Paris.

(14) NASH J. : Le problème de Cauchy pour les équations différentielles d'un fluide général. Bull. Sic. Math. France 90 (1962) 487-497.

(15) NISHIDA T. et MATSUMURA A. : The initial value problem for the equations of motion of viscous and heat concluctive gase . A paraître.

(16) TANI : On the initial value problem for the system of fundamental equations describing the movement of compressible viscous fluid. Preprint.

(17) TEMAM R. : The evolution Navier-Stokes equations - North Holland - Page 427-443.

(18) KLAINERMAN S. et MAJDA A. : Singular limits of quasilinear hyperbolic systems with large parameters and the incompressible limit of compressible fluids - CPAM - vol. XXXIV - 481-524 (1981).

VORTEX THEORY AND EULER AND NAVIER-STOKES EVOLUTION IN TWO DIMENSIONS

C. Marchioro

Dipartimento di Matematica, Università di Trento

38050 Povo (Trento) - Italy

In this note we want to discuss the relation between the evolution of a two dimensional incompressible fluid and the so called "vortex theory". First we introduce such theory in an euristic way.

Consider a nonviscous incompressible fluid (the viscous case will be discussed later). The Euler equation governing the motion reads as :

$$\begin{aligned}
&\frac{\partial}{\partial t}\underline{u}(x,t) + (\underline{u}(x,t)\cdot\nabla)\underline{u}(x,t) = -\nabla p(x,t) \\
&\nabla\cdot\underline{u}(x,t) = 0 \qquad\qquad x\in D\subset R^d \\
&\underline{u}(x,0) = \underline{u}_o(x) \\
&\underline{u}(x,t)\cdot\underline{n} = 0 \quad x\in\partial D \\
&\underline{u}\ \xrightarrow[|x|\to\infty]{}\ 0
\end{aligned} \tag{1}$$

where $\underline{u}(x,t)$ is the velocity, $p(x)$ the pressure and $\underline{n}$ the normal of ∂D. If $d = 2$ (1) becomes :

$$\begin{aligned}
&\frac{\partial}{\partial t} w(x,t) + (\underline{u}(x,t)\cdot\nabla)w(x,t) = 0 \\
&w(x,t) = \text{curl}\,\underline{u}(x,t) = \frac{\partial u_2(x,t)}{\partial x_1} - \frac{\partial u_1(x,t)}{\partial x_2}
\end{aligned} \tag{2.a}$$

$$\underline{u}(x,t) = \int_D \nabla^{\perp} g_D(x,y)\, w(y,t)\,dy$$

$$w(x,0) = w_o(x) \tag{2.b}$$

where $\nabla^{\perp} \equiv \frac{\partial}{\partial x_2}, \frac{\partial}{\partial x_1}$ and $g_D(x,y)$ is the fundamental solution of the Poisson equation with Dirichlet boundary conditions (i.e. $\Delta_x g_D(x,y) = -\delta(x-y)$, $g_D(x,y) = 0$ if x or $y \in \partial D$). If $D \equiv R^2$

$$g(x,y) = -\frac{1}{2} \ln x-y$$

From now on we consider the case $D \equiv R^2$

It is quite natural to write (2) in a weak form

$$\frac{d}{dt} w_t[f] = w_t[(\underline{u} \cdot \nabla) f]$$

$$\underline{u} = \int \nabla_x^{\perp} g(x-y)\, w(y,t)\,dy \tag{3}$$

$$w[f] = \int f(x)\, w(x)\,dx$$

where f is a sufficiently regular function.

Now we want to study the evolution when the vorticity is concentrated in the points $x_i(t)$. It is natural to try to find a solution of the form

$$w_t^N(dx) = \sum_{i=1}^{N} a_i\, \delta_{x_i(t)}(dx) \tag{4}$$

where $a_i \in R$ is called the vortex intensity and $\delta_x(dx)$ denotes the Dirac measure based in x. Obviously inserting eq. (4) in eq. (3), this last does not make sense. This because the velocity is diverging in the points $x_i(t)$ in which "vortices" are localized.

Neglecting this (infinite !) term we obtain

$$\frac{d}{dt} w_t^N[f] = w_t^N [(\underline{u} \cdot \nabla) f]$$

$$\underline{u}(x,t) = \int \nabla_x^{\perp} g(|x-y|) \chi(\{x \neq y\}) w_t^N(dy) \tag{5}$$

where

$$\chi(\{x \neq y\}) = \begin{cases} 1 & \text{if } x \neq y \\ 0 & \text{if } x = y \end{cases}$$

Inserting now eq. (4), by direct computation, we obtain

$$\dot{x}_i^N(t) = \sum_{\substack{j=1 \\ j\neq 1}}^{N} a_j \nabla_{x_i}^{\perp} g(|x_i^N(t) - x_j^N(t)|)$$

$$x_i^N(0) = x_i \tag{6}$$

Thus each vortex moves under a velocity field produced by all other vortices (but not by itself). This is the vortex theory introduced in 19th century by Helmholtz, Kirchhoff and Poincarè.

It is natural to try to find a rigorous connection between this evolution problem (given by N ordinary differential equations) and the Euler equation (that involve infinitely many degree of freedom). This connection is given by the following theorem (not completely correct, as we shall see later).

If $w_o(x)$ is regular enough and w_t is the solution of eq. (3), and $x_i^N(t)$ are the solution of eq. (6) and

$$\sum_{i=1}^{N} a_i \, \delta_{x_i}(dx) \xrightarrow[N \to \infty]{\text{weak}} w_o(x)\,dx \tag{7}$$

then

$$\sum_{i=1}^{N} a_i \, \delta_{x_i(t)}(dx) \xrightarrow[N \to \infty]{\text{weak}} w_t(x)\,dx \tag{8}$$

This theorem, with a small modification which will be precised further, is one of the main results of our research. In effect we need some care in studying eq. (6) because of the singularity of the velocity field generated by each vortex. In fact the global solution of eq. (6) may not exist. To show this, we observe first that the motion is hamiltonian, that it can be written as

$$\begin{aligned} a_i \, \dot{x}_{i1} &= \frac{\partial H}{\partial x_{i2}} \\ a_i \, \dot{x}_{i2} &= - \frac{\partial H}{\partial x_{i1}} \end{aligned} \tag{9}$$

where

$$H = - \frac{1}{4\pi} \sum_{\substack{i,g=1 \\ i \neq j}}^{N} a_i \, a_j \, \ell n \, |x_i - x_j| \tag{10}$$

The energy H does not depend explicitly on the time and so it is a constant of motion. So for $N = 2$ the distance between the two vortices is a constant and the collapse cannot appear. For every N when all a_i have the same sign, the first integrals prevent collapse. In general this is not true; for instance it is possible to see that, already for $N = 3$, there are particular case for which collapses appear:

$$a_1 = 2, \; a_2 = 2, \; a_3 = -1 \; , \; x_1 \equiv (-1,0) \, , \; x_2 \equiv (1,0) \, , \; x_3 \equiv (1,\sqrt{2})$$

eq. (6) reduces (ℓ_{ij} = distance between i,j vortex)

$$\frac{d}{dt} \ell_{ij}^2(t) = - \frac{1}{3\sqrt{2}\pi} \ell_{ij}^2(t) \tag{11}$$

and hence

$$\ell_{ij}(t) = \ell_{ij}(0) \sqrt{1 - \frac{t}{3\sqrt{2}\pi}} \tag{12}$$

Therefore $\ell_{ij}(t) = 0$ if $t = 3\sqrt{2}\pi$ and the r.h.s. of eq. (9) becomes infinite. But these situations are in some sense exceptional. In fact when D is a flat torus (and in eq. (6) g is replaced by g_D) it can be proved that up to an arbitrary time T collapses can happen only for a set of null Lebesgue measure in D^N (1). When $D \subset R^2$ there is also an extra term arising from the interaction of a vortex with the boundary. In this case the same ideas working for $D \equiv T^2$ could be applied combined with natural estimates on the Green function. The proof have been explicitely done when D is a circle. For general bounded regular domain the result is more than reasonable (2). For $D \equiv R^2$ the result is proved when $\sum_{P(N)} a_i \neq 0$ ($P(N)$ subset of the first N integers). In other cases and for unbounded domains no proof is still available.

So we need some cutoff on the vortex interaction and we must remove it when $N \to \infty$. We replace in eq. (6) $g(r)$ by $g_{\varepsilon(N)}(r)$ defined as

$$g_\varepsilon(r) = g(r) \qquad \text{for} \qquad r \geqq \varepsilon$$

and arbitrary extended to an even $C^2(R^1)$ function such that

$$|g'_\varepsilon(r)| \leqq |g'(r)| \quad ; \quad |g''_\varepsilon(r)| \leqq |g''(r)|$$

Theorem 1. Let $w_o(x) \in L_1 \cap L_\infty(R^2)$ and $w_o^N(dx)$ be the measure defined in eq. (4) (with the same positive and negative charge of w_o). For all $w_o^N(dx)$ such that

$$\lim_{N\to\infty} w_o^N[f] = w_o[f] \qquad\qquad (13)$$
$$f \in C^2(R^2) \;, \qquad |\nabla f| \in L_\infty(R^2)$$

and all sequence $\varepsilon = \varepsilon(N)$ such that

$$\lim_{N\to\infty} R(w_o, w_o^N) \exp\{2aL_\varepsilon(T + e^{2aL_\varepsilon T}\} \qquad\qquad (14)$$

where R is a suitable metric generating a topology equivalent to the weak convergence topology, $L_\varepsilon = \max\{\tilde{L}_\varepsilon, 2\max \nabla^\perp g_\varepsilon\}$, $\tilde{L}_\varepsilon$ Lipschitz constant of $\nabla^\perp g_\varepsilon$, $T = \sup t$, $a = \text{const}$.

Then

$$\lim_{N\to\infty} w_t^N[f] = w_t[f] \qquad\qquad (15)$$

where w_t is the solution of the Euler equation.

We do not give here the proof of this theorem. It can be found in Ref. (3). We have stated here the above result when $D \equiv R^2$, but it is easy to obtain the same result when $D \subset R^2$ with smooth enough boundary [(2)].

Finally we note that we have obtained a method to approximate a partial differential equation by a finite dynamical system. This could provide an useful tool for numerical purpose. Of course the speed of the convergence of the method presented here is very small. Stronger hypothesis on the regularity of the initial data and an accurate choice of the cutoff procedure allow us to obtain stronger results.[(4)]

Now we want to obtain a similar result for the Navier-Stokes equation for an incompressible fluid with a spacially homogeneous viscosity.

$$\frac{d}{dt} w_t[f] = w_t [(\underline{u} \cdot \nabla) f] + \nu\, w_t[\Delta f]$$

$$w_o = w \qquad (16)$$

$$\underline{u}(x,t) = \int \nabla^{\perp} g(x-y) w_t(y) dy$$

We prove that this equation can be related with a stochastic vortex model (for rational simplicity we consider now each vortex intensity equal to $\frac{1}{N}$) :

$$d\, x_i^N(t) = \underline{u}_i^{\varepsilon} (\{x_i^N\}) + d\, \underline{b}_i(t) \qquad (17)$$

where $\underline{b}_i(t)$ are N indipendent brownian motion.

$$\underline{u}_i^{\varepsilon} = \frac{1}{N} \sum_{j=1}^{N} \nabla_{x_i} g_{\varepsilon} \left| x_i(t) - x_j^N(t) \right| \qquad (18)$$

That is, each vortex moves in the field produced by other vortices and under the action of N indipendent brownian motions.

We can obtain in this case also a result analogous to theorem 1, using the well known fact that brownian motion is related with the Laplace operator (that is the viscous term). Nevertheless there are some additional difficulties. We want to use the expression

$$\mathbb{E} \left[\frac{1}{N} \sum_{i=1} \delta_{x_i^N(t)}(dx) \right] \qquad (19)$$

($\mathbb{E}$ means expectation) as the good approximation but it is no more a solution of a weak form of the Navier-Stokes equation (neglecting the self energy term) as it happened in the deterministic case. In fact for finite N the expectation cannot be exchanged with the time evolution. When $N \to \infty$ almost every path becomes typical (this phenomenum is called in kinetic theory "propagation of chaos") and we have :

Theorem 2.[3] Let $w \in L_1 \cap L_\infty$ and

$$\frac{1}{N} \sum_{i=1}^{N} \delta_{x_i^N}(dx) \xrightarrow[N\to\infty]{\text{weak}} w\,dx \tag{20}$$

then for all $t > 0$ and a suitable sequence $\varepsilon(N)$

$$\mathbb{E}\left[\frac{1}{N} \sum_{i=1}^{N} \delta_{x_i^N(t)}(dx)\right] \xrightarrow[N\to\infty]{\text{weak}} w_t\,dx \tag{21}$$

where w_t is the solution of the Navier-Stokes equation.

We have proved the result when $D \equiv R^2$. For bounded region the problem is very complicated and not yet solved. In fact the viscosity imposes to the velocity to vanish on the boundery and this condition is not automatically satisfied in the "vortex theory". On the other hand boundaries are very important because they produce the vorticity in the fluid. This production has been numerically investigate by a "vortex method" firstly by Chorin in Ref. (5).

REFERENCES

(1) Dürr D., Pulvirenti M., Commun. Math. Phys. 85, 265 (1982).

(2) Marchioro C., Pulvirenti M., Meeting on "Coulomb Systems and Fluid Dynamics", Trento, May 27 June 10, 1982.

(3) Marchioro C., Pulvirenti M, Commun. Math. Phys., 84, 483 (1982).

(4) Hald O., SIAM J. Numer. Anal. 16 726 (1979).
Beale J.T., Majda A., Math. of Comput., 39, 29 (1982).

(5) Chorin A.J., Fluid Mech., 57, 785 (1973).

FREE BOUNDARY PROBLEMS FOR COMPRESSIBLE VISCOUS FLUIDS.

Alberto VALLI
Dipartimento di Matematica
Università di Trento
38050 POVO , ITALY

1. Statement of the problem.

In this exposition we want to present some results obtained by Secchi-Valli in [4].

Consider the equations of compressible fluid dynamics in their general form (see for instance the paper of Serrin [5])

(1.1) $\rho\,[\,\dot{v} + (v\cdot\nabla)v - f\,] = \operatorname{div} T\,, \qquad (\operatorname{div} T)_i = \Sigma_j D_j T_{ji}\,,$

(1.2) $\dot{\rho} + \operatorname{div}(\rho v) = 0\,,$

(1.3) $\rho\,[\,\dot{e} + v\cdot\nabla e - r\,] = T : D - \operatorname{div} q\,, \qquad T : D = \Sigma_{ij} T_{ij} D_{ij}\,,$

where ρ is the density of the fluid, v the velocity and e the internal energy per unit mass; T is the stress tensor and D is the deformation tensor

$$D_{ij} = \frac{1}{2}\,(\,D_j v_i + D_i v_j\,)\;;$$

q is the heat flux; f and r are the (assigned) external force field per unit mass and the (assigned) heat supply per unit mass per unit time, respectively.

Consider moreover the well known constitutive equations

(1.4) $T_{ij} = [\,-p + (\zeta - \frac{2}{3}\mu)\operatorname{div} v\,]\,\delta_{ij} + 2\,\mu\,D_{ij}$

(Stokesian fluid linearly dependent on D_{ij}) ,

(1.5) $q = -\,\chi\,\nabla\theta$ (Fourier's law),

where p is the pressure, μ and ζ are the viscosity coefficients, θ is the (absolute) temperature and χ is the coefficient of heat conductivity.

We choose as thermodynamic unknowns the density ρ and the temperature θ , and finally consider the following constitutive equations

(1.6) $p = P(\rho,\theta)$,

(1.7) $e = E(\rho,\theta)$,

(1.8) $\mu = \overline{\mu}(\rho,\theta)$,

(1.9) $\zeta = \overline{\zeta}(\rho,\theta)$,

(1.10) $\chi = \overline{\chi}(\rho,\theta)$,

where P, E, $\overline{\mu}$, $\overline{\zeta}$ and $\overline{\chi}$ are known functions subjected to the thermodynamic restrictions (Clausius-Duhem inequalities)

$$(1.11) \qquad \mu \geq 0 , \qquad \zeta \geq 0 , \qquad \chi \geq 0 .$$

Moreover, from the well known relation

$$dS = \frac{1}{\theta}\, dE - \frac{P}{\theta\rho^2}\, d\rho \quad , \qquad S \text{ specific entropy} ,$$

E and P must satisfy the compatibility condition

$$(1.12) \qquad \frac{\partial E}{\partial \rho} = \frac{1}{\rho^2}\left(P - \theta\frac{\partial P}{\partial \theta} \right) .$$

We can now rewrite the equations for the unknowns ρ, v, θ :

$$(1.13) \qquad \rho\, [\dot{v} + (v\cdot\nabla)v - f\,] = -\nabla P + \Sigma_j\, D_j(\bar{\mu} D_j v + \bar{\mu}\nabla v_j) + \nabla\, [(\bar{\zeta} - \tfrac{2}{3}\bar{\mu})\, \operatorname{div} v\,] ,$$

$$(1.14) \qquad \dot{\rho} + v\cdot\nabla\rho + \rho\, \operatorname{div} v = 0 ,$$

$$(1.15) \qquad \rho\, \frac{\partial E}{\partial \theta}\, [\, \dot{\theta} + v\cdot\nabla\theta] = -\theta\, \frac{\partial P}{\partial \theta}\, \operatorname{div} v + \rho r + \operatorname{div}(\bar{\chi}\, \nabla\theta) + (\bar{\zeta} - \tfrac{2}{3}\bar{\mu})(\operatorname{div} v)^2 + \tfrac{1}{2}\bar{\mu}\, \Sigma_{ij}(D_i v_j + D_j v_i)^2 .$$

We want to consider the free boundary problem related to these equations, that is we want to study the motion of a fluid which is not contained in a fixed vessel. The domain $\Omega(t)$ occupied by the fluid is an unknown of the problem, and depends on the internal motion. From the paper of Wehausen-Laitone [8], we know that for this type of problem the following conditions must be satisfied on the boundary of $\Omega(t)$:

(i) Kinematic condition . The boundary of $\Omega(t)$ always consists of the very same material particles.

(ii) Dynamic condition . Suppose that the surface tension of the fluid is zero. Then the stress vector $T \cdot n^t$ (where n^t is the outward unit normal to $\partial\Omega(t)$) must be continuous as one passes through the boundary.

Condition (ii) can be written explicitely in this form

$$(1.16) \qquad [\, -P + (\bar{\zeta} - \tfrac{2}{3}\bar{\mu})\, \operatorname{div} v\,]\, n_i^t + \bar{\mu}\, \Sigma_j (\, D_i v_j + D_j v_i)\, n_j^t = -p_e\, n_i^t ,$$

where p_e is the (assigned) external pressure.

For the temperature θ, we can assigne several boundary conditions, as

(1.17)

(a) Dirichlet condition ;

(b) Neumann condition ;

(c) Heat flux assigned, i.e. $\bar{\chi}\, \frac{\partial \theta}{\partial n} = q_e$ (q_e assigned) ;

(d) Newton's law (third kind condition), i.e.

$$\bar{\chi}\,\frac{\partial\theta}{\partial n} = k\,(\theta_e - \theta)$$ (k given positive constant, θ_e assigned external temperature).

Finally, we consider the following initial conditions

$$(1.18) \qquad v(0) = v_o \quad ,$$

$$(1.19) \qquad \rho(0) = \rho_o \quad ,$$

$$(1.20) \qquad \theta(0) = \theta_o \quad ,$$

$$(1.21) \qquad \Omega(0) = \Omega_o \quad .$$

2. Lagrangian coordinates.

It is usual (and convenient) to rewrite this problem in the Lagrangian formulation, which presents two advantages:

(α) The domain in which the equations must be solved becomes fixed in time.

(β) Condition (i) is automatically satisfied, if we define $\Omega(t)$ as the set occupied at time t by the particles which were in Ω_o at time 0.

More precisely, let $\eta = \eta(t,x)$ be the solution of

$$(2.1) \qquad \begin{cases} \dot{\eta} = v(t,\eta) & \text{in }]0,T[\times \Omega_o \quad , \\ \eta(0) = Id & \text{in } \Omega_o \quad , \end{cases}$$

where Id is the identity function in Ω_o. Define

$$(2.2) \qquad \Omega(t) = \eta(t, \Omega_o) \quad .$$

If η is an homeomorphism, $\eta(t, \partial\Omega_o) = \partial[\eta(t,\Omega_o)] = \partial\Omega(t)$, and condition (i) is satisfied.

On the other hand, anyway, one pays a price: in Lagrangian coordinates the differential operators become strongly non-linear.

3. The equations in the Lagrangian form.

Before rewriting the equations in Lagrangian coordinates, we make some simplifications, in order that the presentation of the paper becomes clearer. However, we recall that it is possible to treat the general case in the same way (see [4] for the precise result). Assume that:

$$f = 0 \; ; \quad \bar{\mu} \text{ constant} > 0 \; , \quad \bar{\zeta} \text{ constant} > 0 \; ;$$

$$P(\rho,\theta) = P(\rho) \text{ (barotropic case)} \; ; \quad \rho_o(x) = 1 \; ; \quad p_e = 0 \; .$$

Observe that in the barotropic case (with constant viscosity coefficients), equation (1.15) is not connected with (1.13) and (1.14). Hence it can be considered separately from the other two.

Set now

$$\bar{v}(t,y) = v(t,\eta(t,y))$$

and analogously for the other unknowns.

The equations become (suppress the bar everywhere, and adopt the sum convention for repeated indices):

$$(3.1)\quad \begin{aligned} &\dot{v}_i - \frac{\mu}{\rho} a_{kj} D_k(a_{sj} D_s v_i + a_{si} D_s v_j) - \frac{1}{\rho}(\zeta - \frac{2}{3}\mu) a_{ki} D_k(a_{sj} D_s v_j) = \\ &= \dot{v}_i + A_i(\rho,\eta)v = -\frac{1}{\rho} a_{ki} D_k P \quad \text{in } Q_T =]0,T[\times \Omega_o , \end{aligned}$$

$$(3.2)\quad \dot{\rho} + \rho\, a_{kj} D_k v_j = 0 \qquad \text{in } Q_T ,$$

$$(3.3)\quad \dot{\eta} = v \qquad \text{in } Q_T ,$$

$$(3.4)\quad \begin{aligned} &\mu\,(a_{sj} D_s v_i + a_{si} D_s v_j) N_j + (\zeta - \frac{2}{3}\mu) a_{sj} D_s v_j N_i = B_i(\eta,N)v = \\ &= P N_i \qquad \text{on }]0,T[\times \partial\Omega_o = \Sigma_T , \end{aligned}$$

$$(3.5)\quad v(0) = v_o \qquad \text{in } \Omega_o ,$$

$$(3.6)\quad \rho(0) = 1 \qquad \text{in } \Omega_o ,$$

$$(3.7)\quad \eta(0) = Id \qquad \text{in } \Omega_o .$$

Here $a_{ji} = a_{ji}(t,x)$ is the entry (j,i) of the matrix $[D\eta]^{-1}$ (and $\{D\eta\}_{ks} = D_s \eta_k$), and $N = N(t,x)$ is the outward unit normal to $\partial[\eta(t, \Omega_o)] = \partial\Omega(t)$ at the point $\eta(t,x)$, that is

$$N(t,x) = n^t(\eta(t,x)) .$$

4. The uniqueness theorem.

Theorem A. Let Ω_o be a bounded and connected open subset of $\mathbb{R}^3$, $\partial\Omega_o \in C^1$, $P \in C^1$. Then the solution of (3.1)-(3.7) is unique in the class of functions (ρ, v, η) such that

$$\rho \geq \hat{\rho} > 0 \ \text{ in } \bar{Q}_T , \quad \det [D\eta] \geq \hat{J} > 0 \ \text{ in } \bar{Q}_T , \quad \eta(t,\cdot) \text{ is injective in } \bar{\Omega}_o \text{ for each } t \in [0,T] \quad ;$$

$$\rho \in L^\infty(Q_T) \quad \text{with} \quad \nabla\rho \in L^2(0,T;L^\infty(\Omega_o)) \quad ;$$

$$v \in L^\infty(Q_T) \quad \text{with} \quad Dv \in L^2(0,T;L^\infty(\Omega_o)) \quad , \quad D^2 v \in L^1(0,T;L^\infty(\Omega_o))$$

$$\eta \in L^\infty(Q_T) \quad \text{with} \quad D\eta \in L^\infty(Q_T) \quad , \quad D^2\eta \in L^2(0,T;L^\infty(\Omega_o)) \ .$$

<u>Proof</u>. By direct computation. Take two solutions (ρ^1, v^1, η^1), (ρ^2, v^2, η^2), and consider the equations satisfied by the difference. Then estimate

$$\frac{1}{2}\frac{d}{dt} \left[\int_{\Omega_o} (J^1 \rho^1 |w|^2 + \gamma^2 + \delta^2 + |D\delta|^2) \right] \quad ,$$

where $J^1 = \det [D\eta^1]$, $w = v^1 - v^2$, $\gamma = \rho^1 - \rho^2$, $\delta = \eta^1 - \eta^2$.

It is easily seen that

$$(4.1) \qquad D_k(a_{kj} J) = 0 \quad \text{in } Q_T \quad , \text{ for each } j = 1,2,3 \ ,$$

and that there exists a suitable scalar function g such that

$$(4.2) \qquad a_{kj} n^o_k J = g N_j \quad \text{on } \Sigma_T \quad , \text{ for each } j = 1,2,3 \ .$$

Moreover one can prove that from Korn's inequality the following estimate holds

$$(4.3) \qquad \begin{aligned} &\int_{\Omega_o} \{ \frac{\mu}{2} \Sigma_{ij} J^1 [a^1_{kj} D_k w_i + a^1_{ki} D_k w_j]^2 + (\zeta - \frac{2}{3}\mu) J^1 a^1_{ki} D_k w_i \cdot \\ &\qquad \cdot a^1_{sj} D_s w_j \} \geqslant \\ &\geqslant \min(\frac{\mu}{2}, \frac{3}{4}\zeta) \int_{\Omega_o} \Sigma_{ij} J^1 [a^1_{kj} D_k w_i + a^1_{ki} D_k w_j]^2 \geqslant \\ &\geqslant c_o \int_{\Omega_o} |Dw|^2 - c'_o \int_{\Omega_o} |w|^2 \qquad ; \end{aligned}$$

(for the proof of the first inequality, see also the following Lemma 6.1).

Then by integrating by parts, taking into account (4.1)-(4.3), one arrives after long calculations to apply Gronwall's lemma, and to obtain the uniqueness. □

As a consequence of this result, one can prove easily an uniqueness theorem for the original problem in Eulerian coordinates.

Finally, it is worthy of noting that this theorem is substantially different from those contained in the well known paper of Serrin [6] , since in our case the motion of $\Omega(t)$ is not known a priori and the boundary conditions are of different type.

5. <u>The existence theorem: strategy</u>.

We want to obtain the existence of a solution by a fixed point argument. We will construct a map Z, defined in a suitable set R_T, by means of a <u>linear</u> parabolic problem, obtained by "freezing" at t = 0 the coefficients of the nonlinear operators in

(3.1), (3.4).

More precisely, we start from v^*, ρ^* and we obtain at first η^* and ρ by solving

$$(5.1)\qquad \begin{cases} \dot{\eta}^* = v & \text{in } Q_T \ , \\ \eta^*(0) = \mathrm{Id} & \text{in } \Omega_o \ , \end{cases}$$

$$(5.2)\qquad \begin{cases} \dot{\rho} = -\rho^* a^*_{kj} D_k v^*_j & \text{in } Q_T \ , \\ \rho(0) = 1 & \text{in } \Omega_o \ , \end{cases}$$

where a^*_{kj} is the entry (k,j) of the matrix $[D\eta^*]^{-1}$ (and analogously in the sequel $N^*(t,x)$ is the outward unit normal to $\eta^*(t, \partial\Omega_o)$ in $\eta^*(t,x)$).

Equations (5.1) and (5.2) are trivially solved by integration, hence we have to concentrate on the problem related to v.

Observe that

$$(5.3)\qquad a^*_{kj}(0,x) = \delta_{kj} \ ,$$

$$(5.4)\qquad N^*_j(0,x) = n^o_j(x) \ ,$$

hence if we "freeze" the coefficient of $A_i(\rho^*,\eta^*)$ in (3.1) and of $B_i(\eta^*,N^*)$ in (3.4) we obtain

$$(5.5)\qquad \begin{cases} \dot{v}_i - \mu D_j(D_i v_j + D_j v_i) - (\zeta - \frac{2}{3}\mu) D_i(\operatorname{div} v) = \dot{v}_i + A_i v = \\ \qquad = F^*_i \qquad \text{in } Q_T \ , \\ \mu (D_i v_j + D_j v_i) n^o_j + (\zeta - \frac{2}{3}\mu) \operatorname{div} v \, n^o_i = B_i v = G^*_i \qquad \text{on } \Sigma_T \ , \\ v(0) = v_o \qquad \text{in } \Omega_o \ , \end{cases}$$

where

$$F^*_i = A_i v^* - A_i(\rho^*,\eta^*) v^* - \frac{1}{\rho^*} a^*_{ki} D_k P^* \ ,$$

$$G^*_i = B_i v^* - B_i(\eta^*,N^*) v^* + P^* N^*_i \ ,$$

and $P^*(t,x) = P(\rho^*(t,x))$.

By solving (5.2) and (5.5) we construct the map $Z : (v^*,\rho^*) \longrightarrow (v,\rho)$, and it is clear that a fixed point of Z is a solution of (3.1)-(3.7).

To obtain this fixed point, we need a "good" existence theorem for the linear problem

$$(5.6)\qquad \begin{cases} \dot{w} + Aw = F & \text{in } Q_T \ , \\ Bw = G & \text{on } \Sigma_T \ , \\ w(0) = v_o & \text{in } \Omega_o \ . \end{cases}$$

6. An existence theorem for linear problem (5.6).

6-i) The homogeneous case: $v_o = 0$, $G = 0$.

Define the operator A in the Sobolev space $H^2(\Omega_o)$ by setting

$$D_A = \{ w \in H^4(\Omega_o) \mid Bw = 0 \text{ on } \partial\Omega_o \} \quad . \tag{6.1}$$

We want to prove that $\lambda + A$ is an isomorphism from D_A (endowed with the graph norm) into $H^2(\Omega_o)$, for $\lambda \in \mathbb{C}$ such that $\mathrm{Re}\,\lambda \geq \lambda_o$.
Moreover we want to prove the estimate

$$\|w\|_{D_A} + |\lambda|^2 \|w\|_o \leq c_1 [\|(A+\lambda)w\|_2 + (1+|\lambda|)\, \|(A+\lambda)w\|_o] , \tag{6.2}$$

for each $\lambda \in \mathbb{C}$ such that $\mathrm{Re}\,\lambda > \lambda_o + 1$, $w \in D_A$.

(Here $\|\cdot\|_k$ is the norm in the Sobolev space $H^k(\Omega_o)$; moreover from now on each constant c_i, C_i will depend at most on the data of the problem).
Consider the bilinear form associated to $A + \lambda$

$$\begin{aligned} a_\lambda(w,u) = & \int_{\Omega_o} \frac{\mu}{2} (D_i w_j + D_j w_i)(D_i \bar{u}_j + D_j \bar{u}_i) + \\ & + \int_{\Omega_o} (\zeta - \frac{2}{3}\mu) \,\mathrm{div}\, w \,\mathrm{div}\, \bar{u} + \lambda \int_{\Omega_o} w\bar{u} \quad . \end{aligned} \tag{6.3}$$

Lemma 6.1. Every bilinear form a_λ is bounded and coercive in $H^1(\Omega_o)$ if $\mathrm{Re}\,\lambda \geq \lambda_o = \min(\frac{\mu}{2}, \frac{3}{4}\zeta)$. The coerciveness constant of each a_λ is equal to λ_o/K (K Korn's constant relative to Ω_o).

Proof. The boundedness is obvious. For the coerciveness, if $\zeta \geq \frac{2}{3}\mu$, one has only to apply Korn's inequality. If $\zeta < \frac{2}{3}\mu$, observe that

$$|\mathrm{div}\, v|^2 \leq 3 \;\Sigma_i |D_i v_i|^2 \quad ,$$

and

$$\begin{aligned} \frac{\mu}{2} \Sigma_{ij} |D_i v_j + D_j v_i|^2 + (\zeta - \frac{2}{3}\mu)\, |\mathrm{div}\, v|^2 = & \frac{\mu}{2} \;\Sigma_i \Sigma_{j\neq i} |D_i v_j + D_j v_i|^2 + \\ & + 3\zeta\, \Sigma_i |D_i v_i|^2 + (2\mu - 3\zeta)\, \Sigma_i |D_i v_i|^2 + (\zeta - \frac{2}{3}\mu)\, |\mathrm{div}\, v|^2 \quad . \end{aligned}$$

Hence from Korn's inequality one proves the result. □

Then we obtain a weak solution of

$$\begin{cases} Aw + \lambda w = \phi \quad , \quad \phi \in L^2(\Omega_o) \quad , \\ Bw = 0 \quad , \end{cases} \tag{6.4}$$

by Lax-Milgram's lemma.

By standard elliptic regularization, one sees that $A + \lambda$ is surjective on $H^2(\Omega_o)$ (observe also that the system

$$A = -\mu\Delta - (\zeta + \frac{1}{3}\mu)\ \nabla\text{div}$$

is strongly elliptic in the sense of Agmon-Douglis-Nirenberg [1], and that the boundary operator B satisfies the complementary condition).

Now we need only to obtain estimate (6.2).

Lemma 6.2. A solution $w \in D_A$ of (6.4) satisfies

$$\|w\|_o \leqslant \frac{c_2}{1+|\lambda|}\|(A+\lambda)w\|_o \quad \text{for each } \lambda \in \mathbb{C},\ \text{Re}\,\lambda > \lambda_o + 1\ . \tag{6.5}$$

Proof. In fact,

$$(A+\lambda)w = (A+\lambda_o)w + (\lambda-\lambda_o)w \quad ,$$

and for each $\lambda \in \mathbb{C}$, $\text{Re}\,\lambda \geqslant \lambda_o$

$$\begin{aligned}|\lambda - \lambda_o|\ \|w\|_o^2 &\leqslant \|(A+\lambda)w\|_o \|w\|_o + |a_{\lambda_o}(w,w)| \leqslant \\ &\leqslant \|(A+\lambda)w\|_o \|w\|_o + c_3\|w\|_1^2 \leqslant \\ &\leqslant \|(A+\lambda)w\|_o \|w\|_o + c_3(K/\lambda_o)\,|a_\lambda(w,w)| \leqslant \\ &\leqslant [1 + c_3(K/\lambda_o)]\ \|(A+\lambda)w\|_o \|w\|_o \quad ,\end{aligned}$$

hence (6.5) holds. □

Lemma 6.3. Estimate (6.2) holds.

Proof. From well known a priori elliptic estimates and from (6.5)

$$\begin{aligned}\|w\|_2 \leqslant c_4\ (\|Aw\|_o + \|w\|_o) &\leqslant c_4\ [\|(A+\lambda)w\|_o + (1+|\lambda|)\ \|w\|_o] \leqslant \\ &\leqslant c_5\|(A+\lambda)w\|_o \ .\end{aligned}$$

Hence

$$\begin{aligned}\|w\|_{D_A} = \|w\|_2 + \|Aw\|_2 &\leqslant \|w\|_2 + \|(A+\lambda)w\|_2 + |\lambda|\,\|w\|_2 \leqslant \\ &\leqslant c_5\ (1+|\lambda|)\ \|(A+\lambda)w\|_o + \|(A+\lambda)w\|_2 \quad ,\end{aligned}$$

$$|\lambda|^2\ \|w\|_o \leqslant c_2\ |\lambda|\ \|(A+\lambda)w\|_o \quad .$$

□

Now one can utilize the Laplace transform to get the solution of (5.6) (see Theorem 5.2, chap. 4 of Lions-Magenes [3]).

Suppose that $F(0,x) = 0$, and extend F by zero for $t < 0$ and by "reflection" for $t > T$.

It is easy to prove that the extension $\tilde{F}$ satisfies

$$(6.6) \qquad \|\tilde{F}\|_{2,1,\mathbb{R}} \leq c_6 \|F\|_{2,1,T} ,$$

where the constant c_6 does not depend on T. (Here the norm $\|\cdot\|_{k,r,T}$ is the norm in $H^{k,r}(Q_T)$, see Lions-Magenes [3] , chap. 4).

Hence, by Theorem 5.2, chap. 4 in [3], we have found a solution w of (5.6) in $\mathbb{R} \times \Omega_o$, such that

$$(6.7) \qquad \|w\|_{4,2,T} \leq c_7 \|\tilde{F}\|_{2,1,\mathbb{R}} \leq c_8 \|F\|_{2,1,T} ,$$

where c_7 and c_8 do not depend on $T \leq T_o$, T_o fixed.

6-ii) The non-homogeneous case.

We have now to solve a trace problem. First of all, one has to extend G to a function $\tilde{G}$ defined in $\mathbb{R} \times \partial\Omega_o$. Take a function Φ defined in $\mathbb{R} \times \partial\Omega_o$ such that $\Phi(0) = G(0)$, and then extend $(G - \Phi)$ to a function $(G - \Phi)^{\sim}$ by zero for $t < 0$ and by "reflection" for $t > T$.

Now one can define

$$(6.8) \qquad \tilde{G} = (G - \Phi)^{\sim} + \Phi ,$$

and it easy to verify that

$$(6.9) \qquad \|\tilde{G}\|_{5/2,5/4;\Sigma_{\mathbb{R}}} \leq c_9 [\|G\|_{5/2,5/4;\Sigma_T} + T^{-1/4} \|G\|_{L^2(\Sigma_T)} + \|G(0)\|_{3/2,\partial\Omega_o}] ,$$

where c_9 does not depend on T. (Here $\|\cdot\|_{r,s;\Sigma_T}$ is the norm in $H^{r,s}(\Sigma_T)$, see Lions-Magenes [3] , chap. 4; moreover, if X is a Banach space and s is not integer, the norm in $H^s(0,T;X)$ is defined in the following way

$$\|w\|^2_{H^s(0,T;X)} = \|w\|^2_{H^{[s]}(0,T;X)} + \int_0^T dt \int_0^T d\tau \frac{\|w(t) - w(\tau)\|^2_X}{|t - \tau|^{1+2(s-[s])}} ,$$

where [s] is the integer part of s).

Furthermore, by following the arguments of Grisvard [2] (adapted in our case to a system of equations), one obtains that, if the (necessary) compatibility conditions

$$(6.10) \qquad \mu\,(D_i v_o^j + D_j v_o^i)\, n_j^o + (\zeta - \tfrac{2}{3}\mu)\, \mathrm{div}\, v_o\, n_i^o = G_i(0) \text{ on } \partial\Omega_o,\ i = 1,2,3$$

are satisfied, one can find a function $W \in H^{4,2}(\mathbb{R} \times \Omega_o)$ such that

$$(6.11) \qquad \begin{cases} BW = \tilde{G} & \text{on } \mathbb{R} \times \partial\Omega_o , \\ W(0) = v_o & \text{in } \Omega_o , \\ \dot{W}(0) = -Av_o + F(0) & \text{in } \Omega_o , \end{cases}$$

and

$$(6.12) \qquad \|W\|_{4,2,\mathbb{R}} \leqslant c_{10} [\|\tilde{G}\|_{5/2,5/4;\Sigma_{\mathbb{R}}} + \|v_o\|_3 + \|F(0)\|_1] .$$

We can find now the solution of

$$(6.13) \qquad \begin{cases} \dot{\Psi} + A\Psi = F - (\dot{W} + AW) & \text{in } \Omega_T , \\ B\Psi = 0 & \text{on } \Sigma_T , \\ \Psi(0) = 0 & \text{in } \Omega_o , \end{cases}$$

since the conditions of the homogeneous case 6-i) are satisfied.

Moreover $w = W + \Psi$ is the solution of (5.6), and from (6.12), (6.9) and (6.7) (applied to problem (6.13)) the following estimate holds

$$(6.14) \qquad \begin{aligned} \|w\|_{4,2,T} \leqslant C_1 [& \|F\|_{2,1,T} + \|G\|_{5/2,5/4;\Sigma_T} + \\ & + T^{-1/4} \|G\|_{L^2(\Sigma_T)} + \|v_o\|_3 + \|F(0)\|_1 + \\ & + \|G(0)\|_{3/2,\partial\Omega_o}] , \end{aligned}$$

where the constant C_1 does not depend on $T \leqslant T_o$, T_o fixed.

7. The existence of a fixed point.

We are now in a condition to apply Schauder's fixed point theorem. First of all, we need to define the (convex and compact) set R_T in which the map Z is defined.

Set

$$(7.1) \qquad E = 2 [\| P(\rho_o)\, n^o \|_{3/2,\partial\Omega_o} + \|\rho_o\|_3 + \|v_o\|_3 + \|Id\|_1] ,$$

and take in the sequel $T \leqslant T_o$, T_o fixed.

Define η to be the solution of

$$\begin{cases} \dot{\eta} = v & \text{in } \Omega_T , \\ \eta(0) = Id & \text{in } \Omega_o ; \end{cases}$$

one easily verifies that there exist constant C_2 and C_3 such that if, for an arbitrary $T \leqslant T_o$, v, ρ and η satisfy $\|v\|_{4,2,T} \leqslant C_1 E$, $|\rho|_{\infty;3;T} \leqslant E$, $\det [D\eta(t,x)] \geqslant 1/2$ in $\bar{\Omega}_T$, $v(0) = v_o$, $\dot{v}(0) = - Av_o$, then

$$(7.2) \qquad | \rho\, a_{kj}\, D_k v_j |_{2;3;T} \leqslant C_2 ,$$

$$(7.3) \qquad | \rho\, a_{kj}\, D_k v_j |_{\infty;2;T} \leqslant C_3 .$$

(Here $|\cdot|_{p;m;T}$ is the norm in $L^p(0,T;H^m(\Omega_o))$, $1 \leqslant p \leqslant \infty$).

Moreover, there exists a constant C_4 such that if, for an arbitrary $T \leqslant T_o$, v, ρ and η satisfy $\|v\|_{4,2,T} \leqslant C_1 E$, $|\rho|_{\infty;3;T} \leqslant E$, $\det [D\eta(t,x)] \geqslant 1/2$ in $\bar{\Omega}_T$, $v(0) = v_o$, $\dot{v}(0) = - Av_o$, and $|\dot{\rho}|_{2;3;T} \leqslant C_2$, then

$$(7.4) \qquad | \partial/\partial t \, [\rho \, a_{kj} \, D_k v_j \,] |_{2;1;T} \leq C_4 \quad .$$

As usual, the constants C_2, C_3 and C_4 depend only on the data of the problem and do not depend on $T \leq T_o$.

Set now

$$R_T = \{ (v,\rho) | \quad \| v \|_{4,2,T} \leq C_1 E, \ v(0) = v_o, \ \dot{v}(0) = - Av_o,$$

$$(7.5) \qquad | \rho |_{\infty;3;T} \leq E, \ |\dot{\rho}|_{2;3;T} \leq C_2, \ |\dot{\rho}|_{\infty;2;T} \leq C_3,$$

$$| \ddot{\rho} |_{2;1;T} \leq C_4, \quad \rho(0) = 1, \quad \rho(t,x) \geq 1/2 \text{ in } \overline{Q}_T \} \ .$$

It is easily seen that $R_T \neq \emptyset$ for each $T > 0$, since from Theorem 2.1, chap. 4 of Lions-Magenes [3] there exists a function $v^{(1)} \in H^{4,2}(\mathbb{R} \times \Omega_o)$ such that $v^{(1)}(0) = v_o$, $\dot{v}^{(1)}(0) = - Av_o$, and $\| v^{(1)} \|_{4,2,\mathbb{R}} \leq C_1 \| v_o \|_3 \leq C_1 (E/2)$. Hence $(v^{(1)},1) \in R_T$ for each $T > 0$.

Take now $(v^*,\rho^*) \in R_T$, and construct η^* as in (5.1). It is trivially verified that for $T \leq T_o$

$$(7.6) \qquad | \eta^* |_{\infty;4;T} \leq c_{11} \quad ;$$

moreover, one can find $T_1 \in]0,T_o]$ such that, for an arbitrary $T \leq T_1$, $(v^*,\rho^*) \in R_T$ implies that

$$(7.7) \qquad \det [D\eta^*(t,x)] \geq 1/2 \quad \text{in } \overline{Q}_T \quad ,$$

$$(7.8) \qquad | \eta^*(t,x) - \eta^*(t,y) | \geq 1/2 \, | x - y | \quad \forall t \in [0,T], \ \forall x,y \in \overline{\Omega}_o \quad .$$

Furthermore, from (7.7) and (7.8), $\eta^*(t,\cdot)$ is a diffeomorphism for each $t \in [0,T]$, $T \leq T_1$; hence the vector field N^* can be expressed on Σ_T in terms of $D\eta^*$.

Let ρ be the solution of (5.2), and let $T \leq T_1$. One easily finds that there exists $T_2 \in]0,T_1]$ such that, for an arbitrary $T \leq T_2$, $(v^*,\rho^*) \in R_T$ implies that

$$\rho(t,x) \geq 1/2 \qquad \text{in } \overline{Q}_T \quad ,$$

and

$$| \rho |_{\infty;3;T} \leq E \quad .$$

Moreover from (7.2), (7.3) and (7.4) $\rho \in R_T$.

We are now in a condition to show that the solution v of (5.5) belongs to R_T, for T small enough. Let $T \leq T_1$; one has to estimate F^* and G^* . By recalling these two (obvious) results

$$(\alpha) \qquad f \in H^1(0,T;X) \implies \sup_{t \in [0,T]} \| f(t) - f(0) \|_X \leq \| \dot{f} \|_{L^2(0,T;X)} \, T^{1/2} \quad ,$$

$$(\beta) \qquad f \in L^\infty(0,T;X) \implies \| f \|_{L^2(0,T;X)} \leq \| f \|_{L^\infty(0,T;X)} \, T^{1/2} \quad ,$$

one obtains after some calculations that

$$\| F^* \|_{2,1,T} \leq c_{12} \, T^{1/2} \quad ,$$

and

$$\| G^* \|_{5/2,5/4;\Sigma_T} \leq c_{13}\, T^{1/2} \quad ;$$

hence from (6.14)

$$(7.9) \qquad \| v \|_{4,2,T} \leq C_1\, [(C_{12}+C_{13})\, T^{1/2} + C_{13}\, T^{1/4} + E/2\,] \quad ;$$

Then there exists $T_3 \in\,]0,T_1]$ such that, for an arbitrary $T \leq T_3$, $(v^*,\rho^*) \in R_T$ implies that

$$\| v \|_{4,2,T} \leq C_1\, E \quad ,$$

i.e. $v \in R_T$.

Set now $T^* = \min\,(T_2,T_3)$. We have proved that the map

$$(7.10) \qquad Z : (v^*,\rho^*) \longrightarrow (v,\rho)$$

satisfies $Z(R_{T^*}) \subset R_{T^*}$. Moreover it is easily seen that R_{T^*} is convex, and that it is compact in

$$(7.11) \qquad X = \{\, (v,\rho) \mid v,\ \rho \in C^0(\,[0,T^*];H^2(\Omega_0))\,\} \quad .$$

Furthermore, by an energy estimate one sees that Z is a continuous map from the topology of X into the topology of $C^0([0,T^*];L^2(\Omega_0))$, hence from X into X by a standard compactness argument.

We can now apply Schauder's fixed point theorem, proving in this way the existence of a fixed point for Z, that is the existence of a solution of (3.1)-(3.7). By Theorem A this fixed point is unique.

More precisely, we have obtained the following theorem:

Theorem B. Let Ω_0 be a bounded and connected open subset of $\mathbb{R}^3$, $\partial\Omega_0 \in C^4$, $P \in C^3$, $v_0 \in H^3(\Omega_0)$. Assume that the (necessary) compatibility conditions (6.10) are satisfied. Then there exist $T^* > 0$, $v \in H^{4,2}(Q_{T^*})$, $\rho \in H^1(0,T^*;H^3(\Omega_0)) \cap H^2(0,T^*;H^1(\Omega_0))$ such that $\rho > 0$ in $\overline{Q}_{T^*}$, and a diffeomorphism $\eta \in H^1(0,T^*;H^4(\Omega_0)) \cap H^3(0,T^*;L^2(\Omega_0))$ such that (v,ρ,η) is a solution of (3.1)-(3.7) in Q_{T^*}.

Since η is a diffeomorphism, one can return to the Eulerian formulation, obtaining in this way a solution of the original free boundary problem. Observe also that, by Sobolev's embedding theorem, $\eta(t,\cdot)$ is a diffeomorphism of class $C^{2+\gamma}$, $\gamma = 1/2$, for each $t \in [0,T^*]$, hence $\partial\Omega(t) = \eta(t,\partial\Omega_0)$ is of class $C^{2+\gamma}$, $\gamma = 1/2$.

Finally, we want to remark that Tani [7] has recently proved an existence (and uniqueness) theorem for a free boundary problem of this type. His proof is obtained by means of successive approximations, by using a different linearization.

References.

[1] S. Agmon - A. Douglis - L. Nirenberg, Estimates near the boundary for solutions of elliptic partial differential equations satisying general boundary conditions, II, Comm. Pure Appl. Math., 17 (1964), 35-92.

[2] P. Grisvard, Caractérisation de quelques espaces d'interpolation, Arch. Rational Mech. Anal., 25 (1967), 40-63.

[3] J.L. Lions - E. Magenes, Problèmes aux limites non homogènes et applications, vol. 2, Dunod, Paris, 1968.

[4] P. Secchi - A. Valli, A free boundary problem for compressible viscous fluids, J. Reine Angew. Math., to appear.

[5] J. Serrin, Mathematical principles of classical fluid mechanics, Handbuch der Physik, Bd. VIII/1, Springer-Verlag, Berlin Göttingen Heidelberg, 1959.

[6] J. Serrin, On the uniqueness of compressible fluid motions, Arch. Rational Mech. Anal., 3 (1959), 271-288.

[7] A. Tani, On the free boundary value problem for compressible viscous fluid motion, J. Math. Kyoto Univ., 21 (1981), 839-859.

[8] J.V. Wehausen - E.V. Laitone, Surface waves, Handbuch der Physik, Bd. IX, 446--778, Springer-Verlag, Berlin Göttingen Heidelberg, 1960.

FONDAZIONE C.I.M.E.
CENTRO INTERNAZIONALE MATEMATICO ESTIVO
INTERNATIONAL MATHEMATICAL SUMMER CENTER

"Theory of Invariants"

is the subject of the First 1982 C.I.M.E. Session.

The Session, sponsored by the Consiglio Nazionale delle Ricerche, will take place under the scientific direction of Prof. FRANCESCO GHERARDELLI (Università di Firenze, Italy), at the Villa «La Querceta», Montecatini Terme, Italy, ***from June 10 to June 18, 1982.***

Courses

a) ***Classical Theory of Invariants and Compactifications of Algebraic Symmetric Spaces.*** (8 lectures in English).
Prof. Corrado DE CONCINI (Università di Pisa, Italy).

b) ***Geometric Invariant Theory and Applications to Moduli Problems.*** (8 lectures in English).
Prof. David GIESEKER (IAS, Princeton, USA).

1. Introduction to geometric invariant theory.
2. Moduli of stable bundles on a smooth curve.
3. Moduli of stable curves.
4. Degeneration techniques in the study of the moduli space of stable vector bundles on a smooth curve.

References:

1. MUMFORD, D., Stability of Projective Varieties. ***L'Enseignement Mathématique XXIII*** fasc. 1-2, 1977.
2. NEWSTEAD, P.E., Lectures on Introduction to Moduli Problems and Orbit Spaces, New Delhi: Narosa Pub. House, 1978 (Tata Lecture Notes).

c) ***Infinite Root Systems, Representations of Quivers and Invariant Theory.*** (8 lectures in English).
Prof. Victor KAC (MIT, Cambridge, USA).

Contents:

Infinite root system and Weyl group.
Rosenlicht theorem, Vinberg's lemma and Weil conjectures.
Quivers and their representations, reflection functors, description of dimensions of indecomposable representations.
Schur representations and the problem of classification of prehomogeneous vector spaces.

References:

1. GABRIEL P., Unzerlegbare Darstellungen, ***Manuscripta Math. 6,*** 71-103 (1972)
2. KAC V.G., Infinite root systems, representations of graphs and invariant theory, ***Inventiones Math. 56,*** 57-92 (1980).
3. RINGEL C.M., Representations of K-species and bimodules, ***J. of Algebra. 41,*** 269-302 (1976).

Seminars

A number of seminars and special lectures will be offered during the Session.

FONDAZIONE C.I.M.E.
CENTRO INTERNAZIONALE MATEMATICO ESTIVO
INTERNATIONAL MATHEMATICAL SUMMER CENTER

"Thermodynamics and Constitutive Equations"

is the subject of the Second 1982 C.I.M.E. Session.

The Session, sponsored by the Consiglio Nazionale delle Ricerche, will take place under the scientific direction of Prof. GIUSEPPE GRIOLI (Università di Padova, Italy) at Noto, Italy, ***from June 23 to July 2, 1982.***

Courses

a) ***Thermodynamcs and Constitutive Relations*** (8 lectures in English).
Prof. Bernard D. COLEMAN (Carnegie-Mellon University, USA).

Lect. 1. Elementary applications of the Clausius-Duhem inequality: theories of elastic and viscous materials, various types of heat conductors, and materials with internal variables.
References:

1. B.D. COLEMAN & W. NOLL, ***Arch. Rational Mech. Anal. 13,*** 167-178 (1963).
2. B.D. COLEMAN & V.J. MIZEL, ***Arch. Rational Mech. Anal. 13,*** 245-261 (1963); ***J. Chem. Phys. 40,*** 1116-1125 (1964).
3. B.D. COLEMAN, ***Proprietà di Media e Teoremi di Confronto in Fisica Matematica,*** C.I.M.E. Session at Bressanone, 1963.
4. B.D. COLEMAN & M.E. GURTIN, ***J. Chem. Phys. 47,*** 597-613 (1967).
5. C. TRUESDELL, ***Rational Thermodynamics,*** McGraw-Hill, New York, (1969)

Lect. 2. Viscoelastic materials and theories of fading memory.
References:

1. B.D. COLEMAN & W. NOLL, ***Arch. Rational Mech. Anal. 6,*** 355-370 (1960); ***Rev. Modern Phys. 33,*** 239-249 (1961).
2. B.D. COLEMAN & V.J. MIZEL, ***Arch. Rational Mech. Anal. 23,*** 87-123 (1966); ***ibid. 29,*** 18-31 (1968); ***ibid. 30,*** 172-196 (1968).
3. B.D. COLEMAN & D.R. OWEN, ***Arch. Rational Mech. Anal. 55,*** 275-299 (1974).

Lect. 3. Thermodynamics of materials with memory.
References:

1. B.D. COLEMAN, ***Arch. Rational Mech. Anal. 17,*** (1964); ***ibid. 17,*** 230-254 (1964).
2. B.D. COLEMAN & V.J. MIZEL, ***Arch. Rational Mech. Anal. 27,*** 255-274 (1967).
3. W.A. DAY, ***The Thermodynamics of Simple Materials with Fading Memory,*** Springer Tracts in Natural Philosophy, Vol. 22, Springer-Verlag, Berlin, (1962).

Lect. 4. Thermodynamics of electromagnetic fields in dissipative media: magnetically active and electrically polarizable materials exhibiting dispersion and absorption, and various non-linear generalizations of them.
References:

1. B.D. COLEMAN & E.H. DILL, ***Arch. Rational Mech. Anal. 51,*** 1-53 (1973); ***Z.A.M.P. 22,*** 691-702 (1971).

Lect. 5. A theory of thermodynamics in which the existence and regularity of entropy and free energy as functions of state are proved rather than assumed, and conditions necessary and sufficient for the uniqueness of these functions can be given.
References:

1. B.D. COLEMAN & D.R. OWEN, ***Arch. Rational Mech. Anal. 54,*** 1-104 (1972).

Lect. 6 & 7. Examples of materials for which the regularity of entropy and free energy functions is not ***a priori*** evident and requires verification.
References:

1. B.D. COLEMAN & D.R. OWEN, ***Arch. Rational Mech. Anal. 59,*** 25-51 (1975); ***ibid. 70,*** 339-354 (1979); ***Rendiconti dell'Accademia Nazionale dei Lincei*** (VIII) ***61,*** 77-81 (1976); ***Annali di Matematica pura ed applicata*** (IV) ***108,*** 189-196 (1976).
2. B.D. COLEMAN, M. FABRIZIO, & D.R. OWEN, On the Thermodynamics of Second Sound in Dielectric Crystals, ***Arch. Rational Mech. Anal.,*** in press.

Lect. 8. Approaches to thermodynamics in which «absolute temperature» is a derived concept.
References:

1. C. TRUESDELL, ***The Tragicomical History of Thermodynamics,*** 1822-1854, New York, Springer-Verlag, (1980)
2. C. TRUESDELL & S. BHARATHA, The Concepts and Logic of Classical Thermodynamics as a Theory of Heat Engines, New York, Springer-Verlag, (1977).
3. J. SERRIN, ***Arch. Rational Mech. Anal. 70,*** 355-371 (1979); ***Lectures on Thermodynamics,*** University of Naples.
4. B.D. COLEMAN, D.R. OWEN, & J. SERRIN, ***Arch. Rational Mech. Anal. 77,*** 103-142 (1982).

b) Title to be communicated.
Prof. C.W. DAFERMOS (Brown University, USA).

c) ***Rational Thermodynamcs of Mixtures*** (8 lectures in English).
Prof. I. MULLER (TU Berlin).

1. Basic Concepts
1.1 Thermodynamic Fields; 1.2 Equations of Balance; 1.3 Constitutive Equations.

2. Thermodynamics
2.1 Entropy Principle; 2.2 Chemical Potentials; 2.3 Diffusion and Thermal Diffusion.

3. Sound Propagation in Mixtures
3.1 The Speed of Diffusion; 3.2 First and Second Sound.

4. Application to Liquid Helium
4.1 Status of Landau's Theory; 4.2 Helium in Rotation.

5. Kinetic Theory
5.1 Boltzmann Equation and Equations of Transfer; 5.2 Maxwellian Iteration; 5.3 Speed of Heat Conduction, Entropy Flux and Material Objectivity.

6. Outlook.

Literature:

On Equation of Balance:
TRUESDELL, C. & TOUPIN, The Chemical Field Theories, Handbuch der Physik III/1, Springer 1960
TRUESDELL, C., Rational Thermodynamics, McGraw-Hill (1969).

On the Entropy Inequality:
MULLER, I., Thermodynamik, Grundlagen der Materialtheorie, Bertelsmann Universitatsverlag, Dusseldorf (1973).
LIU, I-SHIH, Method of Lagrange Multipliers for Exploitation of the Entropy Principle. ***Arch. Rat. Mech. Anal. 40,*** (1972).
MULLER, I., Thermodynamics and Statistical Mechanics of Fluids and Mixtures of Fluids. Lecture Notes of CNR Summer School in Bevi and Canana.

On Liquid Helium:
LANDAU, L.D., The Theory of Superfluidity of Helium II, ***J. Phys.*** (V.S.S.R.) ***5,*** (1941).
LANE, C.T., Superfluid Physics. McGraw-Hill (1962).

On the Kinetic Theory:
CHAPMAN, S. & COWLING, T.G., Mathematical Theory of non-uniform Gases. Cambridge University Press (1961).
MULLER, I., On the Frame Dependence of Stress and Heat Flux, ***Arch. Rat. Mech. Anal. 45,*** (1972).

On the Speed of Heat Conduction:
CATTANEO, C., Sulla conduzione del calore, ***Atti del seminario mat. e fis. Univ. Modena. 3,*** (1948).
MULLER, I., Zur Ausbreitungsgeschwindigkeit von Storungen in kontinuierlichen Medien, Aachener Dissertation (1966).

Seminars

A number of seminars and special lectures will be offered during the Session.

FONDAZIONE C.I.M.E.
CENTRO INTERNAZIONALE MATEMATICO ESTIVO
INTERNATIONAL MATHEMATICAL SUMMER CENTER

"Fluid Dynamics"

is the subject of the Third 1982 C.I.M.E. Session.

The Session, sponsored by the Consiglio Nazionale delle Ricerche, will take place under the scientific direction of Prof. HUGO BEIRAO DA VEIGA (Università di Trento, Italy) at the «Villa Monastero», Varenna, Italy, ***from August 22 to September 1, 1982.***

Courses

a) ***Construction of weak solutions for Hyperbolic Problems*** (6 lectures in English).
Prof. C. BARDOS (Université Paris-Nord, France)

1. Study of a single conservation law and of the Hamilton-Jacobi equation by the viscosity method.
2. Hyperbolic systems. The Rankine Hugoniot condition and the different forms of the Entropy condition.
3. The construction of Di Perna of the solution of a two by two conservation Law in one space variable.
4. Numerical method. The Courant-Friedrichs-Lewy condition and the order of numerical method. Examples of numerical schemas.

Basic references

1. WHITHAM, Linear and non linear waves.
2. COURANT AND HILBERT, Methods of Mathematical Physics.
3. RITCHINGER AND MORTON, Difference Methods of Initial Value Problems.
4. P.L. LIONS, Generalized solutions of the Hamilton Jacobi equation, Pitman.

On the other hand many articles can be used as

1. R.J. DI PERNA, Convergence of approximate solution to conservation law (preprint).
2. P.D. LAX, Shock waves and entropy. In «***Contribution to non linear functional analysis***» (ZARANTONELLO), Academic Press, 603-634 (1971).
3. A. MAJDA and S. OSHER, Numerical viscosity and the entropy condition, ***Comm. Pure Appl. Math. 32,*** 797-838 (1979).
4. S.N. KRUCKOV, First order quasi linear equation with several independent variables, ***Math. U.S.S.R. Sbornik. 10,*** 217-249 (1970).
5. C. DAFERMOS, The entropy rate admissibility criterion for solutions of hyperbolic conservation law, ***J. Diff. Eq.,*** n. 2 (1973).
6. A.Y. LEROUX, A numerical conception of entropy for quasi linear equations.
7. A. HARTER, J.M. HYMAN and P.D. LAX, On finite difference approximation and entropy condition for shocks, ***Comm. Pure Appl. Math. 29,*** 297-322 (1976).

b) ***Compressible Fluids*** (8 lectures in English).
Prof. Andrew MAJDA (University of California, Berkeley, USA)

1. The local existence of smooth solutions to the compressible equations
2. Compressible and incompressible fluids
3. Formation of shock waves in smooth solutions
4. Existence and stability of multidimensional shock fronts.

References

for 1)-2) 1. T. KATO, ***Arch. Rat. Mech. Anal. 58,*** (1975), 181-205
2. S. KLAINERMAN-A. MAIDA, ***C.P.A.M.. 24*** (1981), 481-524
for 3) 3. P.D. LAX, S.I.A.M. Regional Conf. Series, 13
4. S. KLAINERMAN-A. MAJDA, ***C.P.A.M.. 23*** (1980), 241-263
for 4) 5. A. MAJDA, ***Bull. A.M.S.. 4*** (1981), 342-344
3. A. MAJDA, ***Memoirs A.M.S.***, (to appear).

c) ***The concepts of Continuum Thermomechanis*** (8 lectures in English).
Prof. J. SERRIN (University of Minnesota, USA)

1. Thermodynamical structure
2. Laws of thermodynamics

3. Accumulation theorem
4. Continuum thermomechanics and invariance
5. Balance laws of continuum thermomechanics
6. The Clausius-Duhem inequality
7. Example: The Navier-Stokes equations
8. Example: The shock layer in gas dynamics.

The basic literature reference for the subject would be:

1. Mathematical theory of Fluid Dynamics, Handbuch der Physik, Vol. 8/1
2. Notes on Thermodynamics, University of Minnesota, 1981.

Seminars

A number of seminars and special lectures will be offered during the Session.

Vol. 900: P. Deligne, J. S. Milne, A. Ogus, and K.-Y. Shih, Hodge Cycles, Motives, and Shimura Varieties. V, 414 pages. 1982.

Vol. 901: Séminaire Bourbaki vol. 1980/81 Exposés 561–578. III, 299 pages. 1981.

Vol. 902: F. Dumortier, P.R. Rodrigues, and R. Roussarie, Germs of Diffeomorphisms in the Plane. IV, 197 pages. 1981.

Vol. 903: Representations of Algebras. Proceedings, 1980. Edited by M. Auslander and E. Lluis. XV, 371 pages. 1981.

Vol. 904: K. Donner, Extension of Positive Operators and Korovkin Theorems. XII, 182 pages. 1982.

Vol. 905: Differential Geometric Methods in Mathematical Physics. Proceedings, 1980. Edited by H.-D. Doebner, S.J. Andersson, and H.R. Petry. VI, 309 pages. 1982.

Vol. 906: Séminaire de Théorie du Potentiel, Paris, No. 6. Proceedings. Edité par F. Hirsch et G. Mokobodzki. IV, 328 pages. 1982.

Vol. 907: P. Schenzel, Dualisierende Komplexe in der lokalen Algebra und Buchsbaum-Ringe. VII, 161 Seiten. 1982.

Vol. 908: Harmonic Analysis. Proceedings, 1981. Edited by F. Ricci and G. Weiss. V, 325 pages. 1982.

Vol. 909: Numerical Analysis. Proceedings, 1981. Edited by J.P. Hennart. VII, 247 pages. 1982.

Vol. 910: S.S. Abhyankar, Weighted Expansions for Canonical Desingularization. VII, 236 pages. 1982.

Vol. 911: O.G. Jørsboe, L. Mejlbro, The Carleson-Hunt Theorem on Fourier Series. IV, 123 pages. 1982.

Vol. 912: Numerical Analysis. Proceedings, 1981. Edited by G. A. Watson. XIII, 245 pages. 1982.

Vol. 913: O. Tammi, Extremum Problems for Bounded Univalent Functions II. VI, 168 pages. 1982.

Vol. 914: M. L. Warshauer, The Witt Group of Degree k Maps and Asymmetric Inner Product Spaces. IV, 269 pages. 1982.

Vol. 915: Categorical Aspects of Topology and Analysis. Proceedings, 1981. Edited by B. Banaschewski. XI, 385 pages. 1982.

Vol. 916: K.-U. Grusa, Zweidimensionale, interpolierende Lg-Splines und ihre Anwendungen. VIII, 238 Seiten. 1982.

Vol. 917: Brauer Groups in Ring Theory and Algebraic Geometry. Proceedings, 1981. Edited by F. van Oystaeyen and A. Verschoren. VIII, 300 pages. 1982.

Vol. 918: Z. Semadeni, Schauder Bases in Banach Spaces of Continuous Functions. V, 136 pages. 1982.

Vol. 919: Séminaire Pierre Lelong – Henri Skoda (Analyse) Années 1980/81 et Colloque de Wimereux, Mai 1981. Proceedings. Edité par P. Lelong et H. Skoda. VII, 383 pages. 1982.

Vol. 920: Séminaire de Probabilités XVI, 1980/81. Proceedings. Edité par J. Azéma et M. Yor. V, 622 pages. 1982.

Vol. 921: Séminaire de Probabilités XVI, 1980/81. Supplément: Géométrie Différentielle Stochastique. Proceedings. Edité par J. Azéma et M. Yor. III, 285 pages. 1982.

Vol. 922: B. Dacorogna, Weak Continuity and Weak Lower Semicontinuity of Non-Linear Functionals. V, 120 pages. 1982.

Vol. 923: Functional Analysis in Markov Processes. Proceedings, 1981. Edited by M. Fukushima. V, 307 pages. 1982.

Vol. 924: Séminaire d'Algèbre Paul Dubreil et Marie-Paule Malliavin. Proceedings, 1981. Edité par M.-P. Malliavin. V, 461 pages. 1982.

Vol. 925: The Riemann Problem, Complete Integrability and Arithmetic Applications. Proceedings, 1979-1980. Edited by D. Chudnovsky and G. Chudnovsky. VI, 373 pages. 1982.

Vol. 926: Geometric Techniques in Gauge Theories. Proceedings, 1981. Edited by R. Martini and E.M.de Jager. IX, 219 pages. 1982.

Vol. 927: Y. Z. Flicker, The Trace Formula and Base Change for GL (3). XII, 204 pages. 1982.

Vol. 928: Probability Measures on Groups. Proceedings 1981. Edited by H. Heyer. X, 477 pages. 1982.

Vol. 929: Ecole d'Eté de Probabilités de Saint-Flour X – 1980. Proceedings, 1980. Edited by P.L. Hennequin. X, 313 pages. 1982.

Vol. 930: P. Berthelot, L. Breen, et W. Messing, Théorie de Dieudonné Cristalline II. XI, 261 pages. 1982.

Vol. 931: D.M. Arnold, Finite Rank Torsion Free Abelian Groups and Rings. VII, 191 pages. 1982.

Vol. 932: Analytic Theory of Continued Fractions. Proceedings, 1981. Edited by W.B. Jones, W.J. Thron, and H. Waadeland. VI, 240 pages. 1982.

Vol. 933: Lie Algebras and Related Topics. Proceedings, 1981. Edited by D. Winter. VI, 236 pages. 1982.

Vol. 934: M. Sakai, Quadrature Domains. IV, 133 pages. 1982.

Vol. 935: R. Sot, Simple Morphisms in Algebraic Geometry. IV, 146 pages. 1982.

Vol. 936: S.M. Khaleelulla, Counterexamples in Topological Vector Spaces. XXI, 179 pages. 1982.

Vol. 937: E. Combet, Intégrales Exponentielles. VIII, 114 pages. 1982.

Vol. 938: Number Theory. Proceedings, 1981. Edited by K. Alladi. IX, 177 pages. 1982.

Vol. 939: Martingale Theory in Harmonic Analysis and Banach Spaces. Proceedings, 1981. Edited by J.-A. Chao and W.A. Woyczyński. VIII, 225 pages. 1982.

Vol. 940: S. Shelah, Proper Forcing. XXIX, 496 pages. 1982.

Vol. 941: A. Legrand, Homotopie des Espaces de Sections. VII, 132 pages. 1982.

Vol. 942: Theory and Applications of Singular Perturbations. Proceedings, 1981. Edited by W. Eckhaus and E.M. de Jager. V, 363 pages. 1982.

Vol. 943: V. Ancona, G. Tomassini, Modifications Analytiques. IV, 120 pages. 1982.

Vol. 944: Representations of Algebras. Workshop Proceedings, 1980. Edited by M. Auslander and E. Lluis. V, 258 pages. 1982.

Vol. 945: Measure Theory. Oberwolfach 1981, Proceedings. Edited by D. Kölzow and D. Maharam-Stone. XV, 431 pages. 1982.

Vol. 946: N. Spaltenstein, Classes Unipotentes et Sous-groupes de Borel. IX, 259 pages. 1982.

Vol. 947: Algebraic Threefolds. Proceedings, 1981. Edited by A. Conte. VII, 315 pages. 1982.

Vol. 948: Functional Analysis. Proceedings, 1981. Edited by D. Butković, H. Kraljević, and S. Kurepa. X, 239 pages. 1982.

Vol. 949: Harmonic Maps. Proceedings, 1980. Edited by R.J. Knill, M. Kalka and H.C.J. Sealey. V, 158 pages. 1982.

Vol. 950: Complex Analysis. Proceedings, 1980. Edited by J. Eells. IV, 428 pages. 1982.

Vol. 951: Advances in Non-Commutative Ring Theory. Proceedings, 1981. Edited by P.J. Fleury. V, 142 pages. 1982.

Vol. 952: Combinatorial Mathematics IX. Proceedings, 1981. Edited by E. Billington, S. Oates-Williams, and A.P. Street. XI, 443 pages. 1982.

Vol. 953: Iterative Solution of Nonlinear Systems of Equations. Proceedings, 1982. Edited by R. Ansorge, Th. Meis, and W. Törnig. VII, 202 pages. 1982.

Vol. 954: S.G. Pandit, S.G. Deo, Differential Systems Involving Impulses. VII, 102 pages. 1982.

Vol. 955: G. Gierz, Bundles of Topological Vector Spaces and Their Duality. IV, 296 pages. 1982.

Vol. 956: Group Actions and Vector Fields. Proceedings, 1981. Edited by J.B. Carrell. V, 144 pages. 1982.

Vol. 957: Differential Equations. Proceedings, 1981. Edited by D.G. de Figueiredo. VIII, 301 pages. 1982.

Vol. 958: F.R. Beyl, J. Tappe, Group Extensions, Representations, and the Schur Multiplicator. IV, 278 pages. 1982.

Vol. 959: Géométrie Algébrique Réelle et Formes Quadratiques, Proceedings, 1981. Edité par J.-L. Colliot-Thélène, M. Coste, L. Mahé, et M.-F. Roy. X, 458 pages. 1982.

Vol. 960: Multigrid Methods. Proceedings, 1981. Edited by W. Hackbusch and U. Trottenberg. VII, 652 pages. 1982.

Vol. 961: Algebraic Geometry. Proceedings, 1981. Edited by J.M. Aroca, R. Buchweitz, M. Giusti, and M. Merle. X, 500 pages. 1982.

Vol. 962: Category Theory. Proceedings, 1981. Edited by K.H. Kamps, D. Pumplün, and W. Tholen, XV, 322 pages. 1982.

Vol. 963: R. Nottrot, Optimal Processes on Manifolds. VI, 124 pages. 1982.

Vol. 964: Ordinary and Partial Differential Equations. Proceedings, 1982. Edited by W.N. Everitt and B.D. Sleeman. XVIII, 726 pages. 1982.

Vol. 965: Topics in Numerical Analysis. Proceedings, 1981. Edited by P.R. Turner. IX, 202 pages. 1982.

Vol. 966: Algebraic K-Theory. Proceedings, 1980, Part I. Edited by R.K. Dennis. VIII, 407 pages. 1982.

Vol. 967: Algebraic K-Theory. Proceedings, 1980. Part II. VIII, 409 pages. 1982.

Vol. 968: Numerical Integration of Differential Equations and Large Linear Systems. Proceedings, 1980. Edited by J. Hinze. VI, 412 pages. 1982.

Vol. 969: Combinatorial Theory. Proceedings, 1982. Edited by D. Jungnickel and K. Vedder. V, 326 pages. 1982.

Vol. 970: Twistor Geometry and Non-Linear Systems. Proceedings, 1980. Edited by H.-D. Doebner and T.D. Palev. V, 216 pages. 1982.

Vol. 971: Kleinian Groups and Related Topics. Proceedings, 1981. Edited by D.M. Gallo and R.M. Porter. V, 117 pages. 1983.

Vol. 972: Nonlinear Filtering and Stochastic Control. Proceedings, 1981. Edited by S.K. Mitter and A. Moro. VIII, 297 pages. 1983.

Vol. 973: Matrix Pencils. Proceedings, 1982. Edited by B. Kågström and A. Ruhe. XI, 293 pages. 1983.

Vol. 974: A. Draux, Polynômes Orthogonaux Formels – Applications. VI, 625 pages. 1983.

Vol. 975: Radical Banach Algebras and Automatic Continuity. Proceedings, 1981. Edited by J.M. Bachar, W.G. Bade, P.C. Curtis Jr., H.G. Dales and M.P. Thomas. VIII, 470 pages. 1983.

Vol. 976: X. Fernique, P.W. Millar, D.W. Stroock, M. Weber, Ecole d'Eté de Probabilités de Saint-Flour XI – 1981. Edited by P.L. Hennequin. XI, 465 pages. 1983.

Vol. 977: T. Parthasarathy, On Global Univalence Theorems. VIII, 106 pages. 1983.

Vol. 978: J. Ławrynowicz, J. Krzyż, Quasiconformal Mappings in the Plane. VI, 177 pages. 1983.

Vol. 979: Mathematical Theories of Optimization. Proceedings, 1981. Edited by J.P. Cecconi and T. Zolezzi. V, 268 pages. 1983.

Vol. 980: L. Breen. Fonctions thêta et théorème du cube. XIII, 115 pages. 1983.

Vol. 981: Value Distribution Theory. Proceedings, 1981. Edited by I. Laine and S. Rickman. VIII, 245 pages. 1983.

Vol. 982: Stability Problems for Stochastic Models. Proceedings, 1982. Edited by V.V. Kalashnikov and V.M. Zolotarev. XVII, 295 pages. 1983.

Vol. 983: Nonstandard Analysis-Recent Developments. Edited A.E. Hurd. V, 213 pages. 1983.

Vol. 984: A. Bove, J.E. Lewis, C. Parenti, Propagation of Singulariti for Fuchsian Operators. IV, 161 pages. 1983.

Vol. 985: Asymptotic Analysis II. Edited by F. Verhulst. VI, 497 pag 1983.

Vol. 986: Séminaire de Probabilités XVII 1981/82. Proceedin Edited by J. Azéma and M. Yor. V, 512 pages. 1983.

Vol. 987: C.J. Bushnell, A. Fröhlich, Gauss Sums and p-adic Divisi Algebras. XI, 187 pages. 1983.

Vol. 988: J. Schwermer, Kohomologie arithmetisch definierter Gru pen und Eisensteinreihen. III, 170 pages. 1983.

Vol. 989: A.B. Mingarelli, Volterra-Stieltjes Integral Equations a Generalized Ordinary Differential Expressions. XIV, 318 pages. 198

Vol. 990: Probability in Banach Spaces IV. Proceedings, 198 Edited by A. Beck and K. Jacobs. V, 234 pages. 1983.

Vol. 991: Banach Space Theory and its Applications. Proceeding 1981. Edited by A. Pietsch, N. Popa and I. Singer. X, 302 page 1983.

Vol. 992: Harmonic Analysis, Proceedings, 1982. Edited by G. Ma ceri, F. Ricci and G. Weiss. X, 449 pages. 1983.

Vol. 993: R.D. Bourgin, Geometric Aspects of Convex Sets with t Radon-Nikodým Property. XII, 474 pages. 1983.

Vol. 994: J.-L. Journé, Calderón-Zygmund Operators, Pseudo-D ferential Operators and the Cauchy Integral of Calderón. VI, 12 pages. 1983.

Vol. 995: Banach Spaces, Harmonic Analysis, and Probability Theor Proceedings, 1980–1981. Edited by R.C. Blei and S.J. Sidne V, 173 pages. 1983.

Vol. 996: Invariant Theory. Proceedings, 1982. Edited by F. Ghera delli. V, 159 pages. 1983.

Vol. 997: Algebraic Geometry – Open Problems. Edited by C. Ci berto, F. Ghione and F. Orecchia. VIII, 411 pages. 1983.

Vol. 998: Recent Developments in the Algebraic, Analytical, an Topological Theory of Semigroups. Proceedings, 1981. Edited b K.H. Hofmann, H. Jürgensen and H.J. Weinert. VI, 486 pages. 198

Vol. 999: C. Preston, Iterates of Maps on an Interval. VII, 205 page 1983.

Vol. 1000: H. Hopf, Differential Geometry in the Large, VII, 184 page 1983.

Vol. 1001: D.A. Hejhal, The Selberg Trace Formula for PSL(2, IR Volume 2. VIII, 806 pages. 1983.

Vol. 1002: A. Edrei, E.B. Saff, R.S. Varga, Zeros of Sections c Power Series. VIII, 115 pages. 1983.

Vol. 1003: J. Schmets, Spaces of Vector-Valued Continuous Func tions. VI, 117 pages. 1983.

Vol. 1004: Universal Algebra and Lattice Theory. Proceedings, 198 Edited by R.S. Freese and O.C. Garcia. VI, 308 pages. 1983.

Vol. 1005: Numerical Methods. Proceedings, 1982. Edited by V. Pe reyra and A. Reinoza. V, 296 pages. 1983.

Vol. 1006: Abelian Group Theory. Proceedings, 1982/83. Edited b R. Göbel, L. Lady and A. Mader. XVI, 771 pages. 1983.

Vol. 1007: Geometric Dynamics. Proceedings, 1981. Edited by J. Pali Jr. IX, 827 pages. 1983.

Vol. 1008: Algebraic Geometry. Proceedings, 1981. Edited by J. Do gachev. V, 138 pages. 1983.

Vol. 1009: T.A. Chapman, Controlled Simple Homotopy Theory an Applications. III, 94 pages. 1983.

Vol. 1010: J.-E. Dies, Chaînes de Markov sur les permutations. IX 226 pages. 1983.